LIFE OF THE PIGEON

Also by Alexander F. Skutch
(*Illustrated by Dana Gardner)

Life Histories of Central American Birds, 3 vols.
The Quest of the Divine
Life Histories of Central American Highland Birds
The Golden Core of Religion
A Naturalist in Costa Rica
Studies of Tropical American Birds
The Life of the Hummingbird
Parent Birds and Their Young
*A Bird Watcher's Adventures in Tropical America**
Aves de Costa Rica
The Imperative Call
*A Naturalist on a Tropical Farm**
*New Studies of Tropical American Birds**
*Birds of Tropical America**
*Nature through Tropical Windows**
Life Ascending
*Life of the Woodpecker**
*Helpers at Birds' Nests**
*A Naturalist amid Tropical Splendor**
*Life of the Tanager**
Birds Asleep
A Guide to the Birds of Costa Rica (with F. Gary Stiles)*

Life of the Pigeon

BY

Alexander F. Skutch

ILLUSTRATED BY

Dana Gardner

Comstock Publishing Associates a division of

Cornell University Press | Ithaca and London

First published 1991 by Cornell University Press.

Library of Congress Cataloging-in-Publication Data
Skutch, Alexander Frank, 1904–
Life of the pigeon / by Alexander F. Skutch; illustrated by Dana Gardner.
p. cm.
Includes bibliographical references and index.
ISBN 0-8014-2528-X (cloth : alk. paper)
1. Pigeons. I. Title.
QL696.C63S58 1991
598.6'5—dc20 90-55753

Printed in the United States of America.
Color plates printed in Hong Kong.

⊗ The paper in this book meets the minimum requirements of the American National Standard for Information Sciences—Permanence of Paper for Printed Library Materials, ANSI Z39.48–1984.

Contents

Illustrations and Tables

Color plates

Figures

Tables

Preface

Pigeons have probably been familiar to more people for more centuries than have birds of any other family. One or more species are native to nearly all parts of Earth habitable by man. They frequent cultivated fields and gardens as well as wild forests and arid deserts. In addition to indigenous species, domesticated pigeons, derived mostly from the Eurasian Rock Dove, have been carried all over the world, and their feral descendants thrive in many countries. Few cities anywhere lack unconfined pigeons that forage in streets, parks, and gardens, and roost and nest on buildings. Here they are seen, and often fed, by multitudes of people who know few birds of other kinds.

The larger members of the family are mostly called pigeons, the smaller ones doves—names of, respectively, French and Anglo-Saxon origin. The distinction has not been consistently followed, as is evident when we recall that the domesticated and feral races of the large Rock Dove are commonly called pigeons. Pigeons and doves are too similar in structure and behavior to provide a scientific foundation for their separation. By whatever name we know them, these birds have endeared themselves to us by their beauty and responsiveness to our friendly overtures. Mild of aspect, rarely aggressive, good neighbors of other birds, they have become symbols of peace and love, emblems of the goddess Astarte in the ancient Near East and of Aphrodite in the Hellenic world. Nevertheless, these harmless birds are among those most harmed by man.

As a boy, I kept pigeons, not as a source of income but simply because they attracted me strongly. My first major work of carpentry was the construction of a small house, where I lodged beautiful maroon Carneaux, pure white Maltese, and a few common pigeons, all living happily together. Still untrained in scientific observation, I discovered nothing new about my feathered pets. Since coming to Central America, I have neglected no opportunity to learn about the little-known birds that surround me, including the seven species of pigeons that have nested on my farm and nature reservation in the foothills of the Cordillera de Talamanca in southern Costa Rica.

In contrast to the enormous amount of information accumulated and published about the many breeds of pigeons kept by aviculturists, used in scientific experiments, and propagated in huge numbers for the market, knowledge of the habits of the world's free pigeons is surprisingly incomplete. The nests of a large proportion of them have never been found. Aviculturists have supplied valuable information about species that have not been studied in their native lands. Despite the large gaps in our knowledge of free pigeons, enough information has been gathered, by others and myself, to give a comprehensive picture of their lives, as I have attempted to do in this book. Moreover, I have not neglected Domestic Pigeons, whose mysterious ability to find their way home from distant points of release has been freely used by experimenters to throw light on the perplexing problems of avian navigation.

I am indebted to the Western Foundation of Vertebrate Zoology and to Dana Gardner and Robb Reavill for photocopies of articles in scientific journals and chapters in books unavailable in Costa Rica.

Dana Gardner is grateful to the Western Foundation of Vertebrate Zoology, the American Museum of Natural History, and the Los Angeles County Museum of Natural History for the loan of specimens used to illustrate this book.

I give in the index the scientific names of pigeons and other living things mentioned in the text, to avoid repeating them each time these species are mentioned.

ALEXANDER F. SKUTCH

"Los Cusingos,"
San Isidro de El General, Costa Rica

LIFE OF THE PIGEON

1 The Pigeon Family

Fossils of Miocene age reveal that pigeons have inhabited Earth for more than twenty million years. Today, these adaptable birds thrive almost everywhere in the tropical and temperate zones but are absent from the Arctic and Antarctic. By their own efforts, pigeons have reached many remote islands; on others, such as the isolated Hawaiian Islands, where they were until recently absent, they have become established since their introduction by man. Some inhabit wet tropical forests; others live in deserts and savannas; and from lowlands they extend to the high, bleak puna of the Andes and to the Himalayas, where they roost and nest amid perpetual snow. Some authors recognize about 300 species of pigeons and doves; but by lumping similar forms, Derek Goodwin, the most recent monographer of the family, reduced the number to 255.

Although pigeons are found over most of Earth, they are by no means uniformly distributed; by far the greater number of species inhabits the tropics and subtropics. In the New World, Colombia has 32 species; Costa Rica, 24; Panama, 23; Brazil, 21. The United States and Canada together have 8 species not introduced by man, but 6 of them are tropical species that reach north only to the southern fringe of the United States. Since the extinction of the Passenger Pigeon, only 2 species are native to the whole vast area north of this fringe: the Mourning Dove, which ranges from coast to coast, and the Band-tailed Pigeon, which is confined to the West. In the Old World, the British Isles and Germany share the same 5 species, one of them a recent arrival from southeastern Europe; New Guinea and its small coastal islands support 41; Australia supports 22 not introduced by man.

Most pigeons are easily recognized by their rather stout bodies, short necks, small heads, and rounded wings. In size, they range from ground-doves and Diamond Doves 6 to 7 inches (15–18 cm) long to Victoria Crowned Pigeons 33 inches (84 cm) long—from the size of a sparrow to that of a hen Turkey. The little Plain-breasted Ground-Dove weighs slightly over an ounce (about 30 g), the Victoria Crowned Pigeon about 5 pounds (2.3 kg).

Pigeons' bills are mostly short, slender, and pointed. In a few species, including the Thick-billed Ground-Pigeon of New Guinea and the Thick-billed Green Pigeon of southeastern Asia, they are stouter. The bill of the São Thomé Green Pigeon is both thick and hooked. Most peculiar of all is the stout, hooked bill of Samoa's Tooth-billed Pigeon. At the base of pigeons' bills is a fleshy cere, often small; but in the male Seychelles Blue Pigeon it is enlarged, warty, bright red, and conspicuous. The equally prominent cere of the Knob-billed Fruit Dove looks like a big red fruit protruding from the forehead.

Pigeons' eyes are red, orange, yellow, blue, or brown and often have red lids or are set in the midst of bare skin that may be red, purplish, blue, green, white, or gray. The legs and feet, with three forwardly directed toes and one hind toe, as in most perching birds, are often red, purplish red, or pink, less frequently yellow or green.

Pigeons have soft, thick plumage, so loosely attached that it is readily pulled or knocked off, which is perhaps an advantage. It may enable an individual loosely seized by a predator to escape with the loss of a few feathers—much as a lizard may save its life by leaving its readily detachable tail in jaws that have grasped it. Male and female pigeons are often too similar in coloration to be readily distinguishable, although the male's plumage may be slightly more richly colored or iridescent. At the other extreme are pigeons whose sexes differ so greatly that they appear to be different species. The male Blue Ground-Dove has a brownish mate; the male Orange Dove, a green partner. Many pigeons plainly attired in browns or

White-bellied Plumed Pigeon
Petrophassa plumifera
Also called Spinifex Pigeon.
Sexes alike. Arid central
and northern Australia.

grays are beautifully iridescent on the nape and hindneck; or a half-collar, black or white or both together, adorns this region. Other pigeons display green, blue, purple, orange, or red in diverse patterns; a few are largely white, or black with vivid iridescence. Among the few crested species are the Australian White-bellied Plumed Pigeon, which wears a tall, upright spire; the Crested Pigeon, also of Australia, with an almost equally high and slender adornment; and the Crested Long-tailed Pigeon of the Solomon Islands. Loveliest of all are New Guinea's crowned pigeons, with high, lacy crests.

The tails of pigeons are mostly of moderate length, with square, rounded, or sometimes pointed ends. Among the few species with notably long tails are Reinwardt's Long-tailed Pigeon of New Guinea and neighboring islands, the Crested Long-tailed Pigeon already mentioned, the Masked Dove of Africa, the Pin-tailed Green Pigeon of southeastern Asia, and the cuckoo-doves of the genus *Macropygia.*

The forty-three genera in this family, the Columbidae, are classified by Goodwin in four subfamilies. Largest of all is the Columbinae, which comprises the more widespread, familiar pigeons and doves, including all those of the New World and of Europe. Among them are such "typical" pigeons as the Rock Dove and its many domesticated and feral varieties and the Wood Pigeon, the Mourning Dove, and the little Common Ground-Dove. The subfamily also includes quail-doves of the genus *Geotrygon,* which walk, partridgelike, over the ground in tropical American woodlands, and their Old World counterparts of the genus *Gallicolumba,* as likewise the bronzewings and emerald doves (*Phaps, Chalcophaps,* and related genera) of the Australasian region.

Many members of this subfamily glean seeds and small invertebrates from the ground; others subsist mainly on berries and diverse small fruits that they gather in treetops. Many of these pigeons are beautiful in softly blended shades; a few are brilliant. Among the more handsome species in the Old World are the Rock Dove and the largely blue Wood Pigeon and Stock Dove; and in the New World, the strikingly patterned Scaled Pigeon, the male Blue Ground-Dove and its more richly colored relative, the Maroon-chested Ground-Dove, and the Sapphire Quail-Dove. Very different in coloration from most members of the subfamily is the male Emerald Dove, iridescent emerald-green on mantle and wings, brownish purple on neck and breast, with a broad white stripe over the eye. It inhabits wooded country from northern India to eastern Australia.

A strange member of the Columbinae is the large Nicobar Pigeon, which lives on small islands from the eastern Indian Ocean to the Philippines and Solomon Islands. Clad largely in iridescent dark blue and green, it has a short white tail, long wings and legs, and a prominent, knoblike cere. The elongated hackle feathers on its hindneck hang over its back and shoulders

Thick-billed Green Pigeon
Treron curvirostra, male
Nepal and Bengal to Sumatra,
Borneo, and Philippines.

in a glossy mantle. It forages over the ground in deep shade, swallowing fruits and digesting seeds so hard that they would be regurgitated whole by imperial pigeons.

Another odd member is the elegant Pheasant-Pigeon of New Guinea and surrounding islands. Bigger than most pigeons, it walks over the forest floor on long red-and-yellow legs, picking up food with a red bill. Its tail, composed of twenty or twenty-two feathers instead of the twelve or fourteen usual in the family, is laterally compressed, like that of a domestic chicken, rather than spread horizontally. Both sexes have black short-crested heads, glossed with green and blue. The upperparts are iridescent reddish purple and the underparts blackish and richly glossed with purple and blue. The hindneck in one race is a contrasting silvery white. When disturbed on the forest floor, these pigeons rise with loudly beating wings to perch in trees.

The subfamily Treroninae is an assemblage of fruit-eating pigeons of mainly tropical distribution, in India, southeastern Asia, islands of the southwestern Pacific, and Africa south of the Sahara. In their digestive tracts and dietary habits, they fall into two subgroups. The green pigeons of the genus *Treron,* with long, narrow alimentary tracts and grinding gizzards much like those of other seed-eating pigeons, triturate and digest the seeds of the wild figs and other fruits that they ingest, instead of voiding them whole. Thus they are seed predators rather than seed dispersers. Fruit doves of the genus *Ptilinopus* and the larger imperial pigeons of the genus *Ducula* swallow whole fruits that are often surprisingly large, digest

Black-backed Fruit Dove *Ptilinopus cincta*
Sexes similar. Smaller islands of eastern Indonesia.

the pulp in wide guts, and discard the seeds intact and viable, thus serving to disseminate the trees that nourish them. These are among the most lavishly adorned of pigeons, rivaling the parrots in brilliance of attire. The soft green of species of *Treron* is beautifully variegated with orange, yellow, purplish chestnut, or lilac, often with black and yellow on the wings. The Pink-necked Green Pigeon has a pale blue head, bright orange breast, and blue-gray central tail feathers. The more splendid of the fruit doves are too profusely adorned to be described in a few words. One of the more simply patterned is the Magnificent Fruit Dove, or Wompoo Pigeon, of New Guinea and eastern Australia. Both sexes are mainly green tinged with gold. A patch of dark purple, broadening rearward, stretches from throat to abdomen. The legs are green to greenish orange.

The imperial pigeons are named for the impressive size of the larger of them, which much exceeds that of Feral Pigeons. The head, foreparts of the body, and ventral surface of both sexes of the 22-inch (56-cm) Blue-tailed Imperial Pigeon of central Indonesia are pale silvery gray, tinged with pink, with chestnut under tail coverts. The lower back and wings are shining green, shading to dark purplish blue on the tail. The eyes are golden. Very different is the Pied Imperial Pigeon, which roosts and nests on small islands from the Nicobars in the Indian Ocean eastward to the Philippines. Contrasting with its white, yellow-tinctured body are its largely black wings and tail. Other species of this magnificent genus are arrayed in a great diversity of color patterns.

Also in this subfamily are the four blue fruit-eating pigeons of the genus *Alectroenas* of Madagascar and small islands of the western Indian Ocean. One of them, the Mauritius Blue Pigeon, is, alas, extinct; another, the Comoro Blue Pigeon, is endangered. Probably related to the foregoing species in the odd frugivorus Topknot Pigeon of Australia's eastern coast. Its body is silvery and bluish gray, its wings and tail mostly black; on its head, above a thick, laterally compressed bill, it wears a strange double crest, gray on the forehead and chestnut broadly bordered with black above the crown and nape.

The subfamily Didunculinae consists of a single species, the Tooth-billed Pigeon of Samoa. About the size of an average Feral Pigeon, it is dark chestnut on back, rump, and tail. Elsewhere it is predominantly blackish green, glossed with silvery blue on the hindneck. Its strange stout bill, not unlike that of the Dodo, has a hooked upper mandible and three toothlike projections on each side of the lower mandible. As one would expect, with such a bill it eats in a peculiar fashion, which will claim our attention in the following chapter.

The subfamily Gourinae contains only the three species of crowned pigeons, all of New Guinea and nearby small islands. The biggest of all the pigeons, with sixteen tail feathers, they are about the size of a large domestic chicken or a small hen Turkey, and they differ little in plumage. The Blue Crowned Pigeon is mostly medium to dark grayish blue. Its upper back and most wing coverts are dark purplish red, with a conspicuous white patch on the greater coverts. Above its red eyes is a tall, laterally compressed, fan-shaped crest of blue-gray feathers with free barbs, imparting a lacy aspect. The Maroon-breasted Crowned Pigeon has even longer, more filmy crest feathers. Those of the Victoria Crowned Pigeon are tipped with white. Unlike most such magnificent birds, male and female crowned pigeons are alike. They forage on the ground but fly well and rise to the trees to rest when alarmed.

Pigeons are the only extant family of the order Columbiformes. Three

Black-collared Fruit Pigeon
Ducula mullerii
Sexes similar. New Guinea and the Aru Islands.

other pigeonlike birds are now placed in a different family of this order: the Raphidae. The first is the Dodo of tragic fame, once an inhabitant of Mauritius, one of the tiny, isolated Mascarene Islands in the western Indian Ocean. A corpulent terrestrial bird, it stood nearly 4 feet (1.2 m) high, weighed up to about 50 pounds (23 kg), and had a stout hooked beak, short curly tail feathers, and wings too disproportionately small to lift it into the air. Descended from flying ancestors who reached Mauritius over a wide expanse of ocean, it flourished for a long time in its island sanctuary. In the absence of ground predators, it became heavy, flightless, and so devoid of fear that sailors who arrived in the sixteenth century dubbed it "stupid," while callously slaughtering it for food. They brought to the island pigs, cats, and monkeys, which probably hastened the Dodo's extinction by preying on its eggs and chicks. By about 1670 the Dodo had vanished, before anybody knew much about its way of life. All that remain are drawings of living birds that had been shipped to Europe, India, and Japan; a few museum specimens; and many skeletons that have been dug up in caves and swamps on the island.

Thanks to the observations of a Huguenot castaway named Leguat, we know somewhat more about the habits of the Solitaire of the neighboring island of Rodríguez. Although it resembled the Dodo in its large size and flightlessness, it was, nevertheless, so different that some ornithologists have proposed classifying it in a separate family. Males lived with smaller females on defended territories, eating seeds, fruits of the palm *Latania*, and foliage. The female laid a single egg; and the young, after an interval with their parents, appear to have gathered in creches, like those of certain penguins, pelicans, and flamingos. Surviving longer than the Dodo, the Rodríguez Solitaire was last reported seen in about 1760. Less is known about the third species, the Solitaire of the island of Réunion. It appears to have been a solitary inhabitant of remote mountain forests, where it gathered worms and insects from the ground, until its disappearance early in the eighteenth century.

Until recently included in the order Columbiformes, sandgrouse are now considered to be more closely related to plovers. They have been placed in an order all their own, the Pteroclidiformes.

Most pigeons do not migrate, although, after rearing their young, some wander rather widely in search of food, especially in arid regions such as the interior of Australia. Those that nest high in the Himalayas or the southern Andes descend to lower and warmer levels during the boreal or the austral winter. The few truly migratory pigeons rarely undertake journeys comparable in length to those of many other small birds. Mostly, they go no farther than the more southerly parts of their species' breeding range.

One of the more highly migratory species is the Turtle-Dove, which nests in central Asia, in Europe as far north as the British Isles, and in northern Africa. Northern populations travel through the Mediterranean region, the Near East, and Arabia to pass the winter in Africa between the Sahara and the equator. At winter's approach, Eastern, or Rufous, Turtle-Doves, which breed from Siberia and Japan to India and southeastern Asia, move southward from the more northerly portions of this vast area. Similarly, the Red Collared Dove (or Red Turtle-Dove), of much the same distribution, is only a summer resident at higher latitudes.

The Stock Dove and Wood Pigeon withdraw in winter from the parts of their breeding range where winter is harshest but are permanently resident in the British Isles. The hardy northernmost race of the Rock Dove lives throughout the year in the stormy Faeroe Islands at sixty-two degrees north; and a few degrees farther south, in northern Scotland, it has been known to nest in January. Other populations are migratory.

In the Western Hemisphere, the northernmost populations of Mourning Doves migrate farthest south, "leap-frogging" over more southerly populations, occasionally reaching Panama and Hispaniola. However, some individuals brave the winter near the northern limits of their breeding range, in southern Canada and the northernmost area of the contiguous United States, where in severe winters they suffer high mortality and sometimes frozen feet. On being cared for in captivity, some of the unfortunate doves with frozen feet regained the use of their feet and could walk and perch in spite of the loss of the ends of their toes. In the West, Band-tailed Pigeons withdraw from their summer homes in southeastern Alaska, British Columbia, Washington, and Oregon to winter from California and Arizona into Mexico, where the species also breeds. Most of the White-winged Doves of southern Texas and Arizona migrate to Mexico and Central America, but others remain north of the international boundary. After nesting in Canada and the northern half of the United States east of the Plains, vast flocks of the extinct Passenger Pigeon flew southward to winter in the southeastern United States. In South America, where the migration of native-born birds is a less conspicuous phenomenon than on the northern continent, Chilean Pigeons withdraw from the beech (*Nothofagus*) forests of the south to winter in the center of the long, narrow country of Chile. The Australian Pied Imperial Pigeon, known in Australia as the Torres Strait Pigeon, leaves Cape York Peninsula to winter in New Guinea, where the species also nests.

Pigeons appear to migrate chiefly by day. This was certainly true of the Passenger Pigeon, whose immense, compact flocks obscured the sky for hours while they swiftly passed. Mourning Doves also migrate by day, but rarely more than a score together. In Britain, great flocks of migrating Wood Pigeons have been seen and photographed by day; but some travel by night, occasionally striking lighthouses on the east coast. Similarly, Pied Imperial Pigeons are sometimes killed at lighthouses.

Although pigeons are, on the whole, rather sedentary birds, they are from time to time stirred by an urge to disperse, which has driven them over nearly the whole Earth. Early in the present century, the Collared Dove, widespread in southern Asia, was known in Europe only in the far southeast. From this base it spread northwestward over Europe until, in 1952, it was first seen in England. Soon it nested there and spread throughout the British Isles. From China the Collared Dove was introduced into Japan. Without the food it finds in cities, towns, and cultivated fields, this nonmigratory dove could hardly survive the winters of these northern regions.

Free pigeons exposed to the many hazards of northern lands have a short life expectancy. Among those that have survived longest are Rock Doves, Common Ground-Doves, and Inca Doves, which lived for at least 6 years, a Mourning Dove and a Bar-shouldered Dove for 10 years, a Wood Pigeon for 14, and a White-winged Dove for 15. Protected and well attended in aviaries, Mourning Doves have attained ages of 15 and 17 years; White-winged Doves, at least 17 and 25 years; Ringed Turtle-Doves and Domestic Pigeons, more than 30 years. The few available records suggest that the potential longevity of well-sheltered pigeons is slightly greater than that of passerine birds but falls short of that of parrots and many bigger birds.

2 Eating and Drinking

The bills of birds are clues to how they eat. The chisel-like bills of woodpeckers help them to extract insect larvae from trunks and branches of trees. The long, thin bills of hummingbirds probe tubular corollas of flowers, from which the birds suck nectar with their tongues. The thick bills of many finches prepare fruits and seeds for swallowing. The strong bills of parrots, with both mandibles hinged, crush hard seeds or extract tender ones from their coats. The small, slender bills of most pigeons are less specialized. With few exceptions, they serve only to gather from the ground, or to pluck from plants, objects that the pigeons swallow unaltered. The food is stored and softened in the crop, a large, lobed pouch on the side of the esophagus, before it passes onward to the gizzard, the muscular organ in which typical pigeons triturate food with the aid of grit they pick up.

Mainly vegetarian, pigeons forage on the ground or in trees, as occasion requires. Ground-doves, quail-doves, bronzewings, and many others forage largely or wholly on the ground, over which they walk with bobbing heads, gathering seeds of wild plants, grains of cereal crops, fallen berries, and small bulbs. Every morning before sunrise, when I feed our domestic hens on the lawn, up to twenty White-tipped Doves and Gray-chested Doves gather from surrounding pastures and thickets to share the yellow maize. Even Rock Doves, members of a genus that contains many species that forage in trees, find most of their food on open ground.

Probably most ground-feeders mix a few insects and other invertebrates with their grains. Terrestrial snails are frequently eaten, as are termites in warm lands and earthworms on moist soil. Captive Ruddy Quail-Doves eagerly devour mealworms, small slugs and snails, earthworms, and the like, which suggests that they take a substantial proportion of animal food as they hunt over the floor of tropical forests. On the island of Trinidad, this species and the related White-faced Quail-Dove visit the little patches of bare ground which male White-bearded Manakins prepare for their courtship displays, picking up the tiny seeds these small frugivorous birds regurgitate whole after digesting the pulp. Pigeons are not known to scratch or dig with their feet, as gallinaceous birds commonly do, but some

Marquesas Ground-Dove *Gallicolumba rubescens*, male
Marquesas Islands, southern Pacific Ocean.

ground-foragers flick aside loose soil or litter with their bills. Exceptional are Galápagos Doves and strong-beaked Black Bronzewings, who actually dig with their bills, throwing the soil toward themselves as well as sideward. A Galápagos Dove continued for many minutes to dig beneath an opuntia bush, removing an amount of soil equal to its own volume.

Freshly plowed fields, those newly sown with wheat, oats, barley, rice, millet, or sorghum, standing crops of these ripening grains, harvested fields with much spilled grain amid the stubble, and plantings of beans or peas all offer rich rewards not only to habitual ground-feeders but to pigeons that also find much food in trees. Great flocks descend on such fields, where several species may forage together. In compact assemblages birds forage more efficiently because they feel more secure than when alone and spend less time looking up for raptors or other enemies. Among so many pairs of eyes, one bird is almost certain to detect an approaching foe, and by its posture, calls, or rapid flight give timely warning to others. Wood Pigeons foraging on English stubble fields, where much of the time they cannot see one another, appear uneasy. They assume the alarm posture—standing upright with plumage sleeked, tail depressed, and neck outstretched, making the white nape patch very conspicuous—much more frequently than they do when they eat on more open ground, in plain view of many of their companions. In these circumstances, those at the rear of the flock often fly up as though alarmed, pass over others, whose steady feeding seems to reassure them, then descend at the head of the advancing group. When Band-tailed Pigeons in the western United States settle in fields to forage, they advance in similar leap-frog fashion, which is frequent among shorebirds and fishing cormorants and pelicans. Feral Pigeons foraging on the ground peck more frequently when in large than in small flocks.

The more arboreal pigeons gather directly from trees and shrubs the mast of oaks and beeches and, especially in the tropics, a great variety of soft fruits, which they swallow whole. Buds and tender young foliage enrich their diets, which may also include caterpillars and other invertebrates. Fruits of the Puerto Rican royal palm are a preferred food of White-crowned Pigeons. Through much of the year, a pair of Short-billed Pigeons come from the upper levels of the neighboring rain forest to gather the little mucilaginous fruits of mistletoes that infest some small trees in the clearing by our house. They hang head-downward to break pieces from the dangling green fruiting spikes of the cecropia trees so abundant in tropical America, pluck many berries from small trees of the melastome family, and descend still lower to gather the juicy purple berries of the pokeweed (*Phytolacca*) that springs up abundantly in recently burned clearings. On rare occasions, Short-bills drop to the ground to pick up fallen fruits or grit.

In September of a year when the oak trees abundant in the Guatemalan highlands were laden with ripening acorns, Band-tailed Pigeons tried prematurely to pluck them from their cups. Perching precariously near the ends of twigs in the treetops, they grasped the acorns in their bills and struggled strenuously to detach them, usually without success. Often, the heavy birds tugged so hard that they lost their balance and flapped noisily to another branch. A party of a score or more made a great commotion in the treetops.

Six weeks later, when the ground was littered with fallen acorns, the Band-tails still preferred to gather them directly from the trees and swallow them whole, while standing far out on slender twigs. Now that the acorns were easily detached, a flock of Band-tails foraged so silently that I might walk beneath them without becoming aware of their presence until, alarmed by my passage, they all burst out with loudly flapping wings. What strong gizzards they must have, to remove the embryos from their hard shells. With stronger bills and feet capable of holding acorns, the Steller's Jays who shared the bounty with the pigeons removed the shells before swallowing the embryos.

Band-tailed Pigeons also eat mistletoe fruits and small juicy berries of pokeweed, cestrum, and arborescent fuchsias. Like other pigeons of high mountains, they descend to lower altitudes in search of food, perhaps fallen grain in stubble fields. After their meal, they return to the heights, where they roost. As they fly laboriously up long slopes, sometimes against a headwind, they are too easily shot by greedy gunners.

For five years, R. K. Murton and his co-workers studied the foraging of three British pigeons: the Wood Pigeon, the Stock Dove, and the Turtle-Dove. The first of these is the most versatile, eating freely in trees and on the ground and, with its slightly stouter, somewhat hooked bill, tearing leaves and other green vegetation to vary its diet. In winter, Wood Pigeons eat chiefly clover leaves in pastures and harvested grain fields but also the foliage of weeds and, when snow covers the ground, that of turnips, rape, and various other cultivated species of the cabbage family. In spring, they collect grain from newly sown fields of cereals and legumes, and when this is no longer available, they revert to clover, supplemented by tree leaves or buds and flower buds of hawthorn, beech, and ash. By June they are eating large quantities of legumes and leaves of the cabbage family. As cereals ripen, from July to November, they become the Wood Pigeon's mainstay, first from standing crops, later from spilt grain in stubble fields. The birds prefer wheat to barley. From September to November, beech nuts, acorns, and other tree fruits enter increasingly into their fare. When these resources are exhausted, they turn again to clover. Weed seeds, largely from pastures, are collected mainly in May, June, October, and November, but rarely in large amounts. Especially in summer, when the Wood Pigeons feed nestlings, they devour small snails, wood lice, and a few earthworms, but such animal food is never more than a very minor component of their mainly vegetarian diet.

Throughout the year, Stock Doves subsist chiefly on the seeds of weeds, mainly knotgrass (*Polygonum*), chickweed (*Stellaria*), pigweed (*Chenopodium*), and species of the cabbage family, taken from fallow ground, plowed fields, and cereal stubbles; they also eat cereal grains when such are available. They rarely take leaves and buds of trees. The main food of summer-resident Turtle-Doves is seeds of fumitory (*Fumaria*) and grasses, supplemented by cereal grains and capsules of chickweed. Although these three species of pigeons forage largely on farmlands and have become more abundant and widespread with the expansion of agriculture in Britain, their different food preferences prevent serious competition among them.

As mentioned in the preceding chapter, the gorgeous green pigeons have narrow, muscular grinding gizzards and triturate the seeds in fruits, including many figs, that they gather in trees. Fruit doves and imperial pigeons are likewise frugivorous, but they digest only soft pulp in their wide guts and reject the seeds intact, thereby disseminating them in payment for their meals, as do most frugivorous birds, from tanagers to toucans. Foraging fruit doves crawl through their fruit trees in a way that has reminded some observers of parrots. Both fruit doves and imperial pigeons are known to eat fiery red chili peppers (*Capsicum*). With distensible gapes, big imperial pigeons can swallow whole fruits up to about an inch (2.5 cm) in diameter. Several species of imperial pigeons are called "nutmeg pigeons" because they eat wild nutmegs, digesting the aril (the source of the spice called mace) that embraces the seed and probably regurgitating the latter. The Pied Imperial Pigeon and the Australian Pied Imperial Pigeon nest and roost on small islands, from which they fly long distances over the sea each morning for richer food on larger islands or the mainland, returning in the afternoon to the comparative safety of their islets. Despite their size, imperial pigeons can hang inverted to reach fruits.

In the tropics, generously fruiting trees, shrubs, and vines are frequented by a diversity of colorful birds, which rarely try to monopolize the source of their bounty, making a scene delightful to watch. Among the few tropical frugivores known to defend a feeding territory is a Reinwardt's Long-tailed Pigeon of New Guinea who claimed exclusive possession of the berry-laden epiphytic shrub *Schefflera chaetorrachis* and tried to drive away all other visitors, including four species of birds of paradise. Clapping its wings loudly, the big pigeon tried to alight on the intruders.

Tooth-billed Pigeon *Didunculus strigirostris*
Sexes similar. Samoa, central Pacific Ocean.

Among the few pigeons that can prepare food for swallowing are the handsome terrestrial White-bellied Plumed Pigeons, also called Spinifex Pigeons, one of whom, in captivity, was seen to break a peanut by hammering it with its closed bill. Also in an aviary, Marquesas Ground-Doves would jump to seize a grass seed head with their bills, then hold it down with a foot while they plucked out the seeds, much as I have seen Variable Seedeaters do and as has been reported of other finches. Captive Tooth-billed Pigeons nibbled seeds of loquat, almonds, and hemp into fragments, in the manner of a parrot, then ate them. They tore boiled potatoes into large pieces, which they swallowed. With their feet they held down stale bread while they tore it with their bills. Apparently, the best Feral Pigeons can do is to shake a large chunk of bread until a swallowable piece breaks away. Hard or frozen bread defeats their efforts to reduce it to edible fragments. If pigeons, as a family, were as adept at preparing food as are crows, jays, parrots, some finches, and a number of other birds, a larger variety of food might be available to them. However, despite the inability of many species to process their food before they swallow it, pigeons as a rule fare very well.

To help triturate food in the gizzard, pigeons need hard grit, which even highly arboreal species descend to the ground to gather. Wood Pigeons, Red-billed Pigeons, White-crowned Pigeons, Yellow-legged Green Pigeons, Emerald Doves, and others have been seen pecking at salt provided for domestic cattle and eating salt-impregnated soil, which appears to be especially attractive to imperial pigeons.

Although many insectivorous and frugivorous birds obtain from their food much or all of the water they need, granivorous pigeons must drink copiously. They need water not only to moisten and digest their food but also to prepare their crop milk. Kept without water for half an hour after being fed, Domestic Pigeons show obvious signs of distress. Parents who, after drinking deeply, immediately feed young who have begun to take solid food mixed with crop milk make no attempt to do so if they have not had water.

Domestic chickens and many other birds, after taking a little water in their bills, lift their heads to permit it to flow down into their throats; they repeat this act many times when they are very thirsty. Pigeons, however, immerse their bills, sometimes up to their eyes, and, without raising them, suck up all the water they want in one long draft. White-winged Doves may swallow continuously for as much as eight seconds.

Drinking by suction, or perhaps more accurately by a pumping action, was long thought to be peculiar (among birds) to pigeons and sandgrouse, which was one of the reasons for including both in the same order. Now it

Zebra Dove *Geopelia striata*
Also called Peaceful Dove. Sexes alike. Malaya, Indonesia, Australia, Luzon in Philippines. Introduced in Madagascar and Hawaii.

appears that only young sandgrouse drink in this fashion; adults employ the dip-and-tilt method. Moreover, drinking by suction or pumping is not confined to pigeons and juvenile sandgrouse but is practiced by mousebirds, some waxbills, certain shorebirds, the Budgerigar among parrots, and the Australian Bustard. To be able to drink swiftly and continuously is advantageous to birds that must satisfy their thirst in exposed situations, as at the scattered water holes in arid country, where predatory animals (including humans) often take a toll of defenseless creatures driven there by great need.

In deserts and semideserts, pigeons fly long distances, often in large flocks, to the few natural sources of water or to the artificial sources, often underground water pumped up by windmills, that ranchers provide for their cattle. In the arid interior of Australia, Common Bronzewings, widespread in that country, drink chiefly in the dim light after sunset, sometimes continuing until it is nearly dark. Fewer come in the morning twilight; very few are present at the water holes after sunrise, none through most of the day. When they come to drink, a few fly straight to the pool's edge, but most alight a hundred yards (90 m) or more away, delay there for a while, then walk to the water. To the same sources of water, Flock Pigeons come mainly soon after sunrise and in smaller numbers just before sunset. Shy and nervous, in large, dense flocks, they circle and twist in the air, hesitating to swoop down to the water's edge, from which they may rise again in apparent fright before all have had time to drink. Whether some individuals come in both the morning and evening flocks, to drink twice in a day, is not known.

In contrast to these species with definite times for drinking, other pigeons of Australia's dry country, including the Diamond Dove, the Plumed Pigeon, and the Crested Pigeon, visit the water holes through much of the day—the first chiefly in the hot, thirsty middle hours, the third mainly in the forenoon and afternoon. Diamond Doves approach the water with the utmost caution, sometimes perching silently in a nearby tree or shrub for the better part of an hour before they fly down to drink. All these pigeons share the water with many birds of other families and with mammals. Most diurnal birds drink in full daylight; but Bourke's Parrot, like the Common Bronzewing, visits the water holes while the light is dim, even in semidarkness, but chiefly before sunrise instead of after sunset. These birds prefer to assuage their thirst during the hours when diurnal predators, mainly goshawks and falcons, are least likely to assail them. Likewise, in the deserts of the southwestern United States, large flocks of Mourning Doves fly to springs and water holes in the evening twilight, continuing until they become invisible in the dusk.

In the laboratory, Common Ground-Doves with free access to water drank 11.4 percent of their body weight per day; Inca Doves, 14.6 percent; and White-winged Doves, 12.4 percent. After a brief initial decline when deprived of water, White-winged Doves maintained or gained weight while eating the fruits of prickly pear (*Opuntia*) and tomatoes, with a water content of 85 and 95 percent, respectively, as their only sources of water. This suggests that, in the Sonoran Desert of Arizona and northern Mexico, White-winged Doves might satisfy much or all of their water requirement by eating fruits of the giant saguaro cactus, with a water content equal to that of the prickly pear. In southern Arizona, the presence of these doves coincides with the flowering and fruiting season of the giant cactus. The faces and throats of many of the doves are stained red by the moist central pulp of newly opened ripe Saguaro fruits. Some of the seeds embedded in this mass may help to nourish the doves, and the seeds that cling to their plumage or pass undigested through their bodies may be disseminated by them, suggesting a symbiotic relationship.

White-winged Doves find both nutrients and water in the nectar of Saguaro flowers, while effectively pollinating these large, self-sterile blossoms, which expand in the night and remain open through much of the following day. Galápagos Doves eat the pollen and petals of the cactus *Opuntia helleri*, apparently pollinating the flowers. They also dig into the soft, moist pulp of cacti, probably for water as well as for food, which may include larvae and pupae of flies that infest these plants. Galápagos Doves consume many caterpillars as well as a variety of fruits and seeds, flower petals, buds, and leaves.

Inca Doves, smaller and less mobile than the White-winged Doves that live with them in the arid southwestern United States, are not so well adapted to life in the desert far from man but thrive in towns, villages, and cultivated areas where constantly available water permits frequent drinking. Their northward expansion from Mexico, about a hundred years ago, was favored by increased urbanization in southern New Mexico, Arizona, and Texas. Possibly the same is true of their southward spread in the present century from Nicaragua through northwestern Costa Rica to its central plateau.

Pigeons commonly drink while standing beside the water, on the bank of a pool or the side of a trough. White-winged Doves often stand in shallow water while imbibing it. Rock Doves, Domestic Pigeons, and Flock Pigeons occasionally alight on the water. Highly arboreal, frugivorous green pigeons (*Treron*) prefer to drink by sliding down a branch that hangs above the water or projects above it. White-crowned Pigeons in Puerto Rico drink from rainwater collected in the broad axils of the great fronds of royal palms or in tank bromeliads growing on trees. In hot, dry country, pigeons need water not only for moistening their food and secreting crop milk but also for evaporative cooling to avoid overheating. Like other birds, they lack sweat glands and do not obviously perspire, but they lose water through their skin. When they begin to overheat, they open their mouths and rapidly vibrate the skin of the throat to increase evaporation—the process known as "gular flutter." In the southwestern deserts of the United States, Mourning Doves permit their body temperature to rise as high as 113 degrees Fahrenheit (45°C). Since this is slightly warmer than the surrounding air, it enables them to dissipate heat by conduction and radiation.

Recently fledged pigeons, like many other young birds, peck at or pick up almost any small object that attracts their attention by contrasting with its surroundings or by protruding from their perches. Older pigeons may repeatedly pick up and drop an unfamiliar grain, as though testing it, before they finally swallow or reject it. Young pigeons watch their elders eat, often try without success to snatch from an adult's bill an item the latter has just picked up, and after repeated failures help themselves to the food that the other is eating. Adult pigeons learn from experienced companions to eat things with no resemblance to their natural foods, such as the bread that is often provided by their friends in urban parks. A starving pigeon may ignore such bounty unless it sees another individual eating it. Feral Pigeons learn to recognize habitual benefactors, singling them out amid a crowd, flying directly to them, often alighting on them to receive bread or peanuts, one of their favorite foods.

Dana Gardner, the illustrator of this book, told me of a particular instance of the recognition of benefactors by pigeons. While a schoolboy, he had a pet Feral Pigeon named Crystal. Taken from the nest, the bird was raised on a milklike commercial preparation for feeding calves. One day Dana was marching, a mile from home, among about fifty members of his high school band, with a large tuba wrapped around his torso and projecting above his head, when Crystal happened to fly by. Recognizing Dana in the crowd, despite the instrument that partly concealed him, Crystal spiraled out of the sky and landed near his feet. Then, to the amazement of all the musicians, the pigeon continued for several minutes to march beside his master amid the throng.

3 Daily Life

More than most birds, pigeons exhibit the whole range of sociality from solitude to extreme gregariousness. When not engaged in reproduction in monogamous pairs, some live alone, others in immense flocks. The White-tipped Doves that frequent our shady garden, neighboring thickets, and light woods are nearly always alone, at most in pairs, except when they gather to pick up the maize I scatter over the lawn for them in the early morning. The Gray-chested Doves who eat with them are equally unsocial. In the rain forest, I meet Ruddy Quail-Doves walking over the deeply shaded ground more often alone than in pairs, never in larger groups. From the little I have seen and read of other species of quail-doves, they also live alone when not breeding. Even when they nest, these solitary pigeons are seldom seen with a partner, except while they build and, briefly, as they replace each other on the eggs or nestlings. While one incubates or broods, its mate commonly forages beyond view.

In the forest where Ruddy Quail-Doves inhabit the dimly lighted undergrowth, Short-billed Pigeons frequent the brighter canopy. At all seasons, I see these plainly attired pigeons so frequently in pairs that I am confident they remain with their partners throughout the year. In late autumn and early winter in northern California, the minority of Mourning Doves that do not migrate forage in pairs within flocks that spread over stubble fields, much as macaws and amazon parrots in the tropics fly two by two in large flocks. Similarly, in Maryland, some pairs of the Mourning Doves that do not migrate stay together throughout the year, whereas others apparently live separately over the winter, to reunite as the nesting season approaches. At least among permanently resident Mourning Doves, pair bonds tend to endure as long as both members live.

In New Guinea and smaller islands of the southwestern Pacific, a number of fruit doves—Many-colored, Purple-capped, Crimson-capped, and White-bibbed—are usually seen in pairs, sometimes within a flock. In the Mariana Islands, Joseph Marshall, Jr., found White-throated Doves living in pairs in trees, although most species of its genus (*Gallicolumba*) are terrestrial quail-doves. Charles Darwin, who raised every kind of Domestic Pigeon he could obtain, knew that a male and female could easily be mated for life, so different breeds might be kept in the same aviary without hybridizing.

Whether they nest colonially or in scattered pairs, the more social pigeons fly, forage, and roost in companies that may be small or very large. Familiar examples in North America are Mourning Doves, Common Ground-Doves, Inca Doves, White-winged Doves, and Band-tailed Pigeons; in South America, Eared Doves, Ruddy Ground-Doves, Bare-eyed Pigeons, Scaled Pigeons, and Pale-vented Pigeons; in Eurasia, Rock Doves, Wood Pigeons, Snow Pigeons, and Yellow-legged Green Pigeons; in Africa, Olive Pigeons, Ring-necked Doves, Speckled Pigeons, and African Green Pigeons; and in Australasia, Diamond Doves, Zebra Doves, Flock Pigeons, Crested Pigeons, and many imperial pigeons. Most pigeons appear to be social to some degree, the chief exceptions being those inhabiting the undergrowth of heavy forests, dense thickets, and other places where food is dispersed rather than concentrated and where vegetation impedes swift flight from predators.

Especially when disturbed, larger pigeons take flight with loud wing-claps, which are commonly believed to be made by the wings striking together above the back, although R. K. Murton's wide experience with Wood Pigeons and other pigeons led him to conclude that the sound is produced by powerful downstrokes of the wings, like the cracking of a whip. Other pigeons rise into the air with a rattling noise, probably made by the stiff wing quills striking together, or they fly with a whistling sound. Many pigeons fly swiftly, with steadily beating wings, in a straight rather than an

Flock Pigeon *Phaps histrionica*
Two males (left) and female (right).
Interior of Australia.

undulatory course. When crossing a valley from ridge to ridge or a clearing between two tracts of tall forest, pigeons that rest in treetops fly high in the air. A few pigeons have different ways of flying. Diamond Doves close their wings briefly after a series of rapid flaps, thereby losing altitude and tracing an undulatory course, like that of a woodpecker but less pronounced. The big Pied Imperial Pigeon has an "easy almost idling" flight; with several wing-beats often followed by a downward swoop, it "lunges and plunges through the air . . . quite unlike any other wild pigeon" (Harrison, in Smythies 1960).

Pigeons commonly fly at speeds of about 30 to 50 miles (48–80 km) per hour, usually as determined by the speedometer of a car on a parallel course. A Mourning Dove chased by a car achieved 55 miles (88.5 km) per hour; a homing pigeon pursued by a plane traveled 60 miles (97 km) per hour; and another homing pigeon, probably aided by a tail wind, set a record of 94 miles (151 km) per hour. Wood Pigeons migrating above clouds have been known to fly as high as twelve thousand feet (3,660 m) above the ground. Larger pigeons beat their wings at rates varying from 180 to 360 times per minute, according to whether their flight is leisurely or hurried.

Pigeons forage mainly soon after leaving their roosts in the early morning, and again before they retire in the evening, probably with snacks at intervals throughout the day. Inca Doves in the southwestern United States become active earlier on warm, sunny mornings than on cold, overcast days. In New Mexico, a typical flock of ten to twenty Inca Doves starts to eat at about half-past nine in the morning in midwinter; but in southern Texas, the birds begin an hour or more earlier. After foraging for about an hour, they often sun themselves and loaf for perhaps another hour. Alternately foraging and resting, the doves pass their day until they seek their roosts at about four o'clock in the afternoon in New Mexico, about a half-hour later in Texas. In the short days of midwinter in England at fifty-two degrees north latitude, Wood Pigeons, eating mainly clover, forage continuously, with no time for anything else. Not until the long days of June, when grain becomes available and they can satisfy their needs in a few hours, do the majority of these pigeons have time for successful nesting.

Pigeons bathe in rain or standing water. In a shower, a pigeon leans over, with the wing on its upper side raised and its plumage fluffed out, permitting the drops to penetrate the body feathers and wet the wing's undersurface. Then it leans to the opposite side, with the other wing raised, so raindrops can reach other parts of its body. In shallow water where it feels secure, the bird wades in, fluffs out its feathers, makes a few pecking movements in the water, then stands or lies still for a brief interval, often with one wing raised, as in rain-bathing. Then it ducks its head and neck and vigorously beats its partly spread wings in the water, wetting itself much in the manner of a passerine bird. It repeats these movements until it walks or flies from the pool. Then it stands and flaps its wings to shake off the drops before it seeks a safe and preferably sunny spot to dry and arrange its plumage.

Pigeons preen much as other birds do; but unlike many others, they seldom or never appear to anoint their feathers with oil from the preen gland at the base of the tail, which in some species is vestigial or lacking. Apparently, this is because their feathers are waterproofed not by oil but by a fine powder derived from cells that surround the barbules of certain developing feathers known as powder down. Probably this same powder makes it unnecessary for pigeons to indulge in the dust-bathing practiced by many birds. Exceptional in the family are the Bare-faced Ground-Dove of the high central Andes and the Partridge Bronzewing, or Squatter Pigeon, of eastern Australia, which are reported to dust-bathe.

Pigeons enjoy sunning themselves. Seeking a secluded spot in full sunshine, a bird rolls sideward, raises its partly expanded upper wing, and erects all its feathers on the sunward side. It keeps its eyes open but may blink them, as did a little Common Ground-Dove sunbathing for a quarter-hour on a heap of drying magnolia leaves. A number of sociable Inca Doves, arranging themselves symmetrically, sunbathe for from ten seconds to two minutes, then sit upright and preen their body and flight feathers, often quickly repeating this sequence two to four times.

When pigeons forage in flocks, one occasionally tries to displace another

Partridge Bronzewing *Petrophassa scripta*
Also called Squatter Pigeon. Sexes alike. Eastern Australia.

from a source of food. The threatened individual raises one wing, often high above its back, on the side away from the aggressor, as I have frequently witnessed among the White-tipped and Gray-chested doves eating corn on our lawn. In these two doves, gesture displays the wing's beautiful cinnamon underside, the doves' brightest color. Wings, which can deliver resounding blows, are a pigeon's fists, its most effective weapons in defensive or aggressive encounters. The brushes I have seen among feeding doves have rarely led to blows. Once, when I wished to learn the contents of a White-tipped Dove's nest well above my head in a tree, the brooding parent was reluctant to uncover its nestlings. To make it leave the nest so I could view the contents in a mirror attached to a long pole, I gently touched the bird with the stick, whereupon it raised both wings high above its back in a defiant attitude. Then its right wing struck the end of the pole so hard that I feared the dove had injured itself, but its subsequent behavior revealed that both wings were still sound. I have not watched a pigeon defend its nest from a predator, but noises I heard in the dusk while a White-tipped Dove resisted a marauding snake suggested that she struck at it repeatedly with her wings (a story I will tell in more detail in chapter 7).

Among pigeons whose nests are not well dispersed, territorial conflicts arise. As with other birds, in these encounters the invader often appears to be psychologically disadvantaged and is readily repulsed by the displaying territory-holder. If the intruder is not intimidated by aggressive posturing and resists expulsion, the two may fight furiously, pecking with their bills as well as striking with their wings. In such circumstances, Wood Pigeons have been known to struggle almost continuously for as much as two hours. Leaping into the air, face to face, rival White-crowned Pigeons appear to box with their feet. Pigeons' conflicts appear rarely to result in serious injury, for the contestant getting the worst of it can always retreat. The outcome may be quite different when two pigeons who are not mates are confined in a small cage and one attacks the other. If left together long enough, the weaker of the two may finally be pecked to death. The result of such a cruelly unnatural situation has made some people question the propriety of regarding the dove as a symbol of peace.

After their evening meal and perhaps a visit to some source of water, pigeons seek their roosts. Many sleep in trees, often high above the ground, in small or very large groups. In a hamlet in southern Costa Rica, as night approached, I watched about forty Ruddy Ground-Doves fly in from surrounding fields and vanish amid the broad, crowded green leaves of the arborescent *Dracaena fragrans,* where many wintering Baltimore Orioles roosted with them. Other Ruddy Ground-Doves have slept alone, in a dense hedge or amid the dark foliage of an orange tree, not far from their incubating mates. In England, after the breeding season, a population of around four thousand Wood Pigeons, who by day spread over several thousand acres of farmland, slept in two large roosts, each containing up to fifteen hundred birds, while another thousand or more flew farther to join two other large companies of sleeping birds.

In Argentina, highly gregarious Eared Doves sleep in thickets so spiny and densely tangled that a man can hardly penetrate them. Some of these sleeping places are occupied throughout the year, for nesting as well as for roosting. The ground, the stems, and the foliage of the plants become covered with the doves' droppings. At no great distance are permanent watercourses where the birds can drink before they retire. Some fly as much as thirty-seven miles (60 km), at velocities of thirty-one to forty-three miles (50–70 km) per hour, to reach the place where they will rest. For three hours one afternoon, Eared Doves continued to arrive at a dormitory covering 855 acres (350 hectares), until millions had gathered there to sleep. The extinct Passenger Pigeons of the broad-leaved forests of eastern North America probably roosted in even denser concentrations, for it is recorded that they settled on trees in such great masses that thick branches broke under their weight, maiming or killing many when they fell.

Half an hour before sunset in summer, or one or two hours before sunset in midwinter, Inca Doves in New Mexico leave the foraging or resting groups to fly alone, in pairs, or, rarely, in trios to the evergreen trees where they sleep. Here, after the usual shifting about of birds in well-attended roosts, they settle down two or more together, pressed wing to wing. One group of sleepers consisted of seven doves in a compact row, plus two on top. Sometimes, such a cluster of sleeping Inca Doves consists of parents with their fledged young. Fledgling White-tipped Doves roost in trees, near or in contact with both parents, for some nights after they leave their nests.

Not only at night but also in the daytime, Inca Doves in Texas huddle together for warmth when the temperature drops to twenty-one degrees Fahrenheit (−6°C) or lower. They seek a flat surface, such as a windowsill, the roof of a small building, or a wide branch not more than ten feet (3 m) above the ground, in sunshine and sheltered from wind, often near a feeder. Here they form a pyramid, with up to twelve birds in three rows, the longest row at the base, the shortest at the top, all facing the same way with their plumage fluffed out. The doves continually change their positions in the pyramid, those on the outside of the bottom row frequently flying up to the top row, causing a reorganization of the structure. After a few minutes, the birds may again shift their positions. Isolated individuals often rest a few feet from a pyramid. After perhaps an hour, the huddling birds fly down to forage in a flock.

At night, when Inca Doves roost on slender twigs, where they are safer from predators than of flat surfaces, the three-tiered formation appears im-

Inca Dove *Columbina inca*
Sexes nearly alike. Southwestern United States
to central Costa Rica.

practicable. Moreover, on cold nights, the birds save energy by permitting their body temperature to drop, which they could not so readily do if they were densely packed. By such diverse means—pyramiding by day, roosting less compactly, with hypothermy, by night—these very small, nonmigratory doves of tropical origin manage to survive cold winters in more northerly regions that they have only recently colonized. Most of their neighbors—Common Ground-Doves, Mourning Doves, and White-winged Doves—migrate southward to escape winter's chill.

Occasionally, pigeons sleep on the ground. Although Mourning Doves usually roost in trees, some populations prefer the ground in closely grazed pastures, recently mowed hayfields, or other open spaces with sparse vegetation, near where they nest or forage. Here they rest singly or in loose groups of up to a dozen. On an August night in California, forty-two Mourning Doves were flushed from a sixty-acre (24-hectare) pasture. On winter nights in this region, the doves often rest on or beside unpaved rural roads. When disturbed in the night, they commonly fly almost straight up to a height of about forty feet (12 m), then descend with rapidly fluttering wings, their legs extended as though reaching blindly for the ground, reminding one of helicopters about to land.

In dry north-central Australia, wholly terrestrial White-bellied Plumed Pigeons forage, nest, and sleep on stony ground. They rest, often close together, in slight depressions in bare sand or amid pebbles, on gentle slopes or near hilltops, nearly always on the lee side of some small, flat-topped shrub, large rock, or clump of spinifex. They do not sleep beneath the shrub or in contact with the large rock or tussock of grass that shelters them, apparently because it might impede instant flight from some nocturnal prowler. Coveys of Bare-eyed Partridge Bronzewings sleep on the ground under thick cover.

In the bleak winters of the Faeroe Islands, Rock Doves sleep communally in holes and crevices in the rocky escarpments where in summer they nest, much as feral individuals of their species roost in sheltered spots on urban buildings. In another harsh environment, the Peruvian puna around thirteen thousand feet (4,000 m) above sea level, Black-winged Ground-Doves, which are grayish brown with much black on wings and tail, take advantage of a very different shelter. Over the wide barren landscapes of the high plateau stand, singly or loosely clumped, plants of a gigantic member of the pineapple family, *Puya raimondii*. From the summit of a thick, leafy trunk up to twelve feet (4 m) high springs a single huge inflorescence about twenty feet (6 m) tall, densely covered with innumerable greenish white flowers. The long, broad, crowded leaves of this bromeliad bear along their margins formidable inwardly curved, sharp spines. Since no other plant of the treeless puna offers birds such good protection from predators and the rigors of the climate, a number of species seek it for nesting, sleeping, and shelter on inclement days.

The Black-winged Ground-Doves forage in flocks of twenty or more on grassy ground around the puyas and nest and sleep on its leaves. To pass nights that can be bitterly cold, they push as far inward toward the stem as they can and rest facing outward, their bodies parallel to the leaves, sometimes as many as thirty on a single plant. Sheltered from rain and hail, they occupy the same niche night after night, until thick deposits of their droppings cover the grooved leaves. Unfortunately, the puya plant is not wholly friendly to its lodgers; the doves, as well as smaller birds, get caught on the sharp, inwardly directed marginal hooks, which penetrate more deeply into their flesh the more they struggle to extricate themselves. Probably this tragedy occurs most frequently when the birds are frightened by attacking predators, some of whom also leave their desiccated bodies impaled on the spines. On the same high Peruvian plateau, Bare-faced Ground-Doves avoid this hazard by lodging in holes in the walls of buildings.

To find birds roosting in the dense vegetation of humid tropical forests is difficult (and not devoid of danger from venomous snakes that hunt actively after nightfall). One night, while I prowled through lowland forest in Panama, searching for sleeping birds, my flashlight's beam picked out one of the very few roosting birds I have found in tropical woodland, not counting those that lodge in holes in trees or in dormitory nests that can be located in the daytime and watched as the birds retire in the evening. The sleeping bird rested on a slender horizontal branch about ten feet (3 m) above the ground, in a fairly clear space amid tangled growth. After waking, it remained calmly on its perch in the flashlight's rays, while I scrutinized it from all sides to confirm its identity. Satisfied that it was a Gray-chested Dove, I turned off the light and stole away in the darkness. This was a memorable encounter, for I have found no other pigeon roosting in the underwood of tropical forests.

To discover pigeons sleeping in the canopy of high tropical forests is even more improbable, but exceptions occur. Fruit doves roosting in the upper levels of Oriental and Australian forests excrete the indigestible seeds of fruits they have swallowed whole during the day in such numbers that, plunking noisily against the foliage of the undergrowth, the falling seeds call attention to the doves perching far above.

While sleeping, the Gray-chested Dove whom I found had its head exposed, with forwardly directed bill. As far as I can learn, whether roosting in a tree or sitting on a nest, pigeons always sleep with the head exposed, sometimes with the bill resting against the chest, instead of with the head turned back and snuggled in the feathers of a shoulder, as many other birds sleep. In this, pigeons resemble hummingbirds, who also sleep with the head exposed.

4 Voice and Courtship

Pigeons are not among the most vocally gifted of birds. Their notes have a distinctive quality that makes them easy to recognize; one rarely confuses their voices with those of other birds. Mostly, pigeons are said to coo, an imitative word. Despite a family resemblance, however, the vocalizations of pigeons vary greatly in tone, cadence, and volume. Frequently, they sound solemn or even melancholy, rather than lighthearted or joyous, as is evident from the names popularly applied to certain members of the family: North America has a Mourning Dove, Africa a Mourning Collared Dove. The low-pitched notes of certain doves have been called moans. The deep, full, far-carrying *cooo-cu-cooo* of the Scaled Pigeon has earned it the appellation *paloma tora* ("bull pigeon") in Venezuela. As befits their size, the big imperial pigeons utter barking, booming, and grunting notes, loud and deep. By inflating the skin of the neck or chest, pigeons impart resonance to their notes, which many of them deliver with bills almost or quite closed. They do not spare their voices; some repeat simple phrases so persistently through the day that people find them monotonous or annoying. Nearly always, however, the voices of pigeons are pleasant and comforting, suggestive of rural or sylvan tranquillity and nature's gentler aspects. After a lapse of many decades, I can still hear in my mind the sweetly pensive song of the Mourning Dove which sounded in the woods and fields that I roamed in my boyhood.

Although we may not think of the cooing of pigeons as singing, functionally many of the utterances are songs, for, no less than the more musical performances of songbirds, they announce possession of a breeding territory and attract a mate. On rare occasions, pigeons' verses are so elaborate and prolonged that they have been spontaneously called "songs" by country people. Among the more accomplished singers is the White-winged Dove, known also as the Singing Dove and, in Central America, El Cantorix. The male's song may last ten to twenty seconds. The doves I heard in the highlands of Guatemala seemed to sing *cuu-cu-c'c' cu cuu c'cu cuuu.* Alexander Wetmore transcribed the White-wing's song somewhat differently: *who hoo, whooo hoo, hoo-ah, hoo-hoo-áh, who-oo.* Whether the difference is to be ascribed to the singers or the auditors, I do not know;

White-winged Dove *Zenaida asiatica*
Sexes alike. Southwestern United States to western Panama; Pacific coast of South America from southwestern Ecuador to northern Chile; Bahamas; Greater Antilles.

only rarely do two people paraphrase the utterances of birds in just the same way.

Another pigeon with an elaborate song is the Marianas Fruit Dove of Guam and neighboring islands in the mid-Pacific, which all day long repeats, *cooo, cooo, cooo, cu-cu-cucucu cu-cu-cu coo coo coo.* Instead of making the typical cooing sounds, the green pigeons of southern Asia, Indonesia, and Africa proclaim their presence with whistles and flutelike notes. Those of the Wedge-tailed Green Pigeon have won high praise for richness, beauty, and a musical quality not unlike that of the singing human voice.

To repeat their advertising song or call, many pigeons, including the Scaled, Band-tailed, and Red-billed, perch in high treetops. Surprisingly, Blue Ground-Doves, which forage on the ground and build nests at no great height, deliver their soft *coot coot* from an exposed branch high in a tree. White-tipped Doves prefer lower, less exposed stations, often amid the dense foliage of an orange or mango tree or in a thicket. The Ruddy Quail-Dove utters its low, moaning *cooo* in the deep shade of the rain forest, rarely more than head high. Pigeons may call chiefly in the morning, through a sunny day, or in late afternoon, as has been reported of certain populations of the Band-tailed Pigeon.

Although pigeons' vocabularies are limited, their advertising songs are varied; and most appear to have different notes for diverse occasions. From the White-tipped Doves who are singing while I write, I most often hear a moaning *coo-coo* in an unmistakable tone. At other times they sing *coo-woo* or *cu cu cu coo-ruuu,* higher in pitch and more musical, almost soprano in tone. Sometimes these versions are alternated. When addressing a potential mate, the male dove swells out his neck and calls *cucuroo.* Richard F. Johnston distinguished four vocalizations of Inca Doves. Their song is a simple *coo-coo,* often interminably reiterated. In an aggressive mood, they call *cut-cut-ca-doah.* Their courtship call differs only in emphasis. When alarmed, they repeat, at intervals of a second or two, a soft *cut,* which, like the corresponding notes of many other birds, is difficult to trace to its source.

To Hans Peeters, the advertising song of the Band-tailed Pigeon sounded like *who͞o-ho͞o–who͞o-ho͞o–who͞o-hóo-who͞o-ro͞o–who͞o-ro͞o,* weak and low, with little carrying power. In display flight, Band-tails emit chirping notes, not unlike the sound made by running a finger slowly over the teeth of a comb, and audible at a much greater distance than is the advertising coo. When closely pressed by another while it forages, a Band-tail protests with nasal grunts, the lowest pitched of all its notes. Males slightly alarmed by a person approaching their nests stretch their necks forward and peer about while emitting a short, soft *whoo.* When their mates arrive unusually late to relieve their long spells on the eggs, incubating males greet—or scold?—them with a short, low-pitched *croo.* Nestling Band-tails beg with thin piping notes that do not carry far. The vocalizations of female pigeons are often similar to those of their mates, but weaker and less frequently heard.

Many kinds of pigeons adjust well to life in aviaries, with the result that more has been learned about the courtship of undomesticated as well as long-domesticated species in these artificial situations than in their natural habitats. However, one can never be certain that a captive bird is behaving just as it would if unconfined. One of the most thorough studies of the courtship and mating of free pigeons is that of Wood Pigeons in England by R. K. Murton and A. J. Isaacson and reported in Murton's book *The Wood Pigeon.* Since in its main features the courtship of this species differs little from that of many others, let us take it as our paradigm, then notice certain differences in other pigeons.

Like many other birds, pigeons claim territories, which in crowded colonies may be small. The first step in reproduction is usually the acquisition of such a defended space. Rarely in February, most often in March, male Wood Pigeons begin to establish their territories. Early in the morning, the more advanced of them fly down from the treetops, where they roost communally, to the midcanopy, where they repeat the *coo-cooo-cooo co co* that announces possession and warns others that the space will be defended. Other males, who will nest apart, in hedgerows or copses, also sing in them at this time. When the main flock flies from the roost to the fields where the doves forage, all these males accompany them, or follow soon afterward, leaving the woods deserted for most of the day. In the evening, when the flock returns to roost, the same males go first to their territories, to sing, display, and, if necessary, attack intruders before they join the main company for the night. As the days grow longer and warmer, more males come into breeding condition and claim territories. Soon, each sleeps in his own chosen space. Likewise, as food becomes more plentiful and he needs less time to find it, he passes more of the day in his territory, cooing, displaying, chasing intruders, fighting those that refuse to leave.

Display flights are widespread, but not universal, in the pigeon family. At intervals, the territorial male Wood Pigeon flies up above his treetop, stalls high in the air, claps his wings loudly, glides forward on set wings, losing altitude, flaps upward to regain it, claps again, then volplanes down into another treetop, having traced an undulatory course. The function of the wing-clapping flights is not clear. Although often performed above the territory, the flight is not confined to that space; it may occur when the dove is leaving the area, when he is driven from the nest, or while he is passing over the foraging ground. Sometimes the pigeon claps to a human intruder. Evidently the display flight serves, according to circumstances, to advertise possession of territory, to attract a mate, or to warn intruding males or predators.

As with many other birds whose sexes look alike, a Wood Pigeon must reveal its sex to a potential mate by its behavior. An unmated female who wanders into a male's territory is greeted by a bowing display, which, with variations, appears to be practiced by nearly all pigeons. The male leans far forward, bill pointing downward, tail raised and partly fanned, neck inflated with feathers erected, displaying on each side a white patch surrounded by iridescent pinkish purple and green feathers. Simultaneously, he contracts the pupil of each eye, dilating its golden iris. He calls *co roo co co co coo,* much in the manner of a Domestic Pigeon. Greeted by this exhibition of the pigeon's splendor in an attitude suggestive of appeasement rather than belligerence, another male usually retreats. A female seeking a partner returns again and again, gradually permitting the male to approach closer as she becomes more confident. Perhaps uncertain of the sex or intention of the new arrival, the male may alternate threat postures with bowing; and a timid female may also assume a threatening attitude. Finally, her fear subsiding, she remains with the male as his mate. As with many other birds, she has won her right to stay in his territory by her pacific refusal to be evicted.

In the mating season, Wood Pigeons frequently give the bowing display at a distance from their territories. One individual, foraging in wheat stubble amid a hundred others, bowed successively to sixteen of them, and some of those so addressed displayed in turn to others. If the bird addressed did not fly away, it dismissed the presumptuous one with a peck or a flick of the wing. Probably some of the individuals thus accosted by a male were of his own sex or females already paired. Occasionally, the object of the display responded favorably, and the two flew off together, the displaying bird following the other. Thus, without a territory, a male may win a mate, who accompanies him in quest of one, sometimes fighting on his side when he tries to push in among established pairs. Similarly, when I scatter corn over the grass as the nesting season approaches in February, a White-tipped Dove sometimes crouches low and calls *cu cu cu cu cucuuu,* while he faces another, who is nearly always more interested in picking up the grains than in viewing the performance.

Although some kinds of birds choose a nest site together, often the male pigeon takes the lead in this search. With a territory and a partner, the male Wood Pigeon begins the activity known as nest-calling. He sits in an appropriate site, perhaps already with the rudiments of a nest, and pecks downward with his bill, at the same time vibrating his wings and twitching his raised tail, while he calls *oooo-aaarh.* His attitude and movements, suggestive of actual nest-building, encourage his partner to engage in this occupation. Females nest-call less frequently than males, with much softer notes.

After a male and female have paired, they strengthen the bond and prepare themselves for the work ahead by various intimate interactions. The first of these is caressing, which is often difficult to distinguish from mutual preening. One day, as a female Wood Pigeon approached an eggless nest where a male had been nest-calling, he increased the tempo of his notes and vigorously nodded his head. Standing beside him, she also bowed her head and uttered a much softer and lower nest-call. As the male's excitement diminished, she started to nibble the feathers of his head and neck, concentrating her attention on those around his eyes. Although caressing is usually seen at the nest site or empty nest, where a pair are most often together, it may occur almost anywhere. Frequently, the mated pigeons preen each other reciprocally, giving attention principally to parts not accessible to a bird's own bill, such as the head and neck. They may at times remove vermin from the plumage, but this is clearly not the main function of the activity, which is affective rather than hygienic.

Caressing often leads to billing, an activity so characteristic of pigeons that it has given birth to the expression "billing and cooing." Most frequently at the instigation of the female, the partners hold and fondle each other's bills. Sometimes the male opens his mouth and the female inserts her closed bill. The regurgitative movements that follow may be only symbolic, but sometimes the male feeds his partner, much as he might feed a nestling. Occasionally, the female makes similar movements while their bills are together.

Caressing, billing, and courtship feeding are frequently the prelude to coition, which may be followed by preening by both sexes. One day I watched a male Blue Ground-Dove strut about on a horizontal branch at the edge of a thicket, turning rapidly to face now one side, now the opposite side. When a brown female alighted close beside him, he took her bill in his and appeared to feed her. Dropping her bill, he mounted her briefly, then stepped down on the other side of the branch and seemed to feed her again. Once more he stood on her back, after which he got down and took her bill in his from the left side, as at first. After these three feedings, actual or symbolic, she flew off and left him alone.

As with many other birds, male pigeons guard their mates most closely as the date of laying approaches. The behavior of the jealous male, called "driving" by pigeon fanciers, is most frequently seen when the birds are crowded, as in an aviary, or where many free individuals gather at a concentrated source of food. Apparently, the male drives his consort to remove her from the vicinity of other males, not, as is commonly believed, to make her return to her nest. Wherever she goes, he walks close behind her, perhaps pecking her gently, but sometimes, especially if she approaches too near a male whom he fears, striking her roughly. If she flies, he follows her closely through the air. When he has driven her far enough from potential rivals, he may move ahead and lead the way. A less frequent motive for driving by the male is to remove his partner from a strange place where he feels insecure or from near a human he does not wholly trust, although he does not feel sufficiently threatened to take instant flight.

We have followed the male Wood Pigeon while he establishes his territory, wins a mate by the bowing display, attracts her to an appropriate site by nest-calling, and becomes more intimate with her by caressing, mutual preening, billing, and courtship feeding, until the pair are attuned closely enough to undertake the shared tasks that lie ahead. This sequence is basically the same in other pigeons whose courtship has been carefully followed, but some exhibit interesting variations. Let us look briefly at a few of them before we survey the fascinating process of completing the nest, incubating the eggs, and rearing the young.

A displaying Band-tailed Pigeon glides straight outward and slightly downward from a high treetop, gaining momentum. Then, swerving to either right or left, he traces three partly overlapping circles in the air, each from fifty to two hundred yards (45–180 m) in diameter. He covers about three-quarters of each circle by a long glide, the remaining quarter with rapid, shallow wing-beats. Instead of clapping his wings, he emits the chirping sound already described. In the late afternoons, after a group of Band-tails had returned from foraging to their nesting grove, Peeters watched five and sometimes seven circling at the same time, the sight of one bird flying in circles having incited his neighbors to do likewise. These flights appeared to be territorial rather than nuptial.

Stock Doves also display communally, tracing wide horizontal loops by alternately gliding and slowly beating their wings, to the accompaniment of clapping. With noisy flapping, a male Mourning Dove rises from a tree obliquely upward to a height of a hundred feet (30 m) or more, then on widely spread, motionless wings glides back to Earth in one or more sweeping curves. After a few minutes, he may repeat this spectacular performance. The male Barred Cuckoo-Dove of southeastern Asia flies up from a treetop with clapping wings to a height of about fifty feet (15 m), then spirals back to his perch with wings spread and body feathers erected, especially those of the rump. Similar display flights are given by female pigeons, but they tend to be shorter and less frequent than those of males. Often the female's flight is set off by that of her mate.

In the half-century that I have lived surrounded by terrestrial White-tipped and Gray-chested doves, I have never seen one of them display by flying high with clapping wings. If any of the other five species of pigeons

Crested Pigeon *Ocyphaps lophotes*
Sexes alike. Most of Australia.

which have nested in our garden or in the adjoining forest in southern Costa Rica give such displays, they are certainly rare. The same appears to be true of Inca Doves in drier regions. Early in the mating season, male Inca Doves bob their heads at other individuals of both sexes. Males almost never bob in return; females do, thereby revealing their sex.

Another behavior that helps sex recognition, in these Inca Doves without sexual differences in plumage, is mounting. Early in the season, before pairs are formed, males often stand on the backs of other members of their flock. Another male actively dislodges the rider, whereas females usually ignore him, often continuing to forage while he tries to maintain his balance. After associating in this fashion for a few seconds, the two may preen each other for a minute or two, without forging a lasting bond. Later, mounting without coition may lead to the formation of pairs.

The Inca Dove's most spectacular courtship display is vertical tail-fanning. With body horizontal, a male stands a foot or two in front of a female, facing her. He raises his tail straight upward and spreads its feathers sideward, exhibiting a bold black-and-white pattern above his prominently scaled back. After a pair has started to nest, head-bobbing and mutual preening continue to reinforce the bond between them until the end of the breeding season.

The bowing displays frequent in the courtship of pigeons are modified to reveal most effectively the outstanding adornments of each species. On the ground, a male crowned pigeon (*Goura*) steps briskly in circles around his partner. While she stands motionless, he bows so deeply that his head is inverted, his bill points toward his breast, and his lovely, lacy headdress sweeps the ground at his feet. He raises his spread tail, partly opens his wings to expose the white patch on each, and accompanies his vigorous bows with low-pitched notes: *boom-pa, boom-pa boom-pa.*

The male White-bellied Plumed Pigeon bows with bill directed vertically downward and tall, pointed crest rising with a slight forward tilt. He raises and partly opens his wings to show the iridescent green-and-purple patches on their coverts and erects and spreads his tail to display the black outer feathers. His pupils contract while the irides expand to make his golden eyes stand out vividly from the bare red skin around them in his piebald black, white, and blue-gray face. He does not always direct this striking display to his mate or another pigeon; standing alone on a rock or some other slight elevation, he may bend forward and coo to the world at large.

In his equivalent of the bowing display, the male Luzon Bleeding-Heart runs swiftly after another pigeon, then suddenly stops, facing it. Depressing his tail, he rears upward and puffs out his pink-tinged white breast to display more prominently the blood red patch, which looks like a bleeding wound, in its center. Then, cooing, he inclines his head forward to slightly below its usual position. A bowing Common Bronzewing half spreads his wings and tilts them forward to exhibit the brilliant iridescent patches on their coverts, arranged in a barred pattern. After addressing a female, he may offer her his open bill, into which, after an answering bow, she inserts hers. Next, he passes his neck over hers and presses down, to which she may respond by stretching her neck over his. Apparently rare among pigeons, similar neck caresses were practiced by courting Passenger Pigeons. In his bowing display, a Masked Dove moves both wings slightly and rhythmically at the rate of about one hundred times a minute.

Pigeons, like a number of other birds, seem to lack innate recognition of their own species; if it is present, it is so weak that it is readily overridden by experiences in early life. If a young pigeon is reared by pigeons of a different species or by humans, it becomes imprinted on its foster parents and thenceforth prefers them as social or sexual companions. Barbary Doves most readily become attached to a human fosterer at the age of seven to nine days, when their fear responses are beginning to appear but have not yet become strong. The artificially induced preference for a social or sexual companion of a different species or variety may be attenuated or eradicated by prolonged subsequent association with other individuals of the pigeon's own kind. Under natural conditions, pigeons (and other birds) are nearly always so closely associated with parents, and often other individuals, of their own kind that they prefer this kind to all others.

Domestic and Feral pigeons, which exhibit great diversity of color patterns, may show definite preferences. Nancy Burley found that both sexes of confined Feral Pigeons preferred to mate with "blues" rather than "ash reds." Although males chose "blue bars" as frequently as "blue checkers," females were more often attracted by the checkers. Highly experienced individuals, who would make the most efficient parents, were preferred by both sexes over birds with less experience in breeding. However, both sexes discriminated against individuals more than seven years old, an age at which, despite increased reproductive experience, parental capability appears to wane.

Masked Dove *Oena capensis,* male
Also called Namaqua Dove. Central and southern Africa, Madagascar, southwestern Arabian Peninsula, Socotra.

5 Nests and Eggs

Pigeons have not developed the art of nest-building to a high degree. They neither weave, like New World orioles and their relatives and Old World weavers, nor use cobwebs to bind their structures together, as many small birds do. Their nests are simple, open constructions, at best shallow bowls, often hardly more than platforms, so loosely put together that they need a firm foundation. Most pigeons build their nests above the ground, from a foot (30 cm) or so up to the canopy of lofty rain forests, in trees, shrubs, or vine tangles. Occasionally they choose the abandoned nest of some other bird to support their own. In Iowa, one-fifth of all Mourning Doves' nests are built on old nests of other birds, chiefly the substantial structures of American Robins and Common Grackles, which often remain intact over winter. Galápagos Doves frequently occupy empty nests of Galápagos Mockingbirds.

In humid forests of tropical America, Ruddy Quail-Doves place their slight nests on the tops of old stumps, on the firm horizontal leaves of an epiphytic fern or aroid growing on a tree trunk, or on a large fallen palm frond that has lodged horizontally in a shrub or tangle of vines, rarely as high as nine feet (2.7 m) above the ground. Often, these quail-doves cover their foundation with barely enough material to keep their eggs from touching it or rolling off. Despite their lackadaisical building, often on precarious supports, these doves have firmly established themselves over an immense territory, from Mexico to Paraguay.

A few pigeons occupy holes. The Stock Dove nests in cavities in trees (often those carved by woodpeckers), in crannies or on sheltered ledges of cliffs or deserted buildings, and occasionally in rabbit burrows or on the ground beneath dense bushes. Like other birds that breed in holes they cannot make for themselves, Stock Doves find their preferred nest sites in short supply and may fight for them. In lieu of a suitable cavity, they may lay their eggs in an abandoned Wood Pigeon's nest. Another occupant of tree holes is the Black Wood Pigeon of subtropical islands of the Japanese archipelago. Occasionally, Australia's Common Bronzewing nests in a hollow in a tree.

Caves, crannies, and sheltered ledges in cliffs are the preferred nest sites

Snow Pigeon *Columba leuconota*
Sexes alike. Himalayas, Tibet, and mountains of western China.

of a number of species of *Columba* and a few other pigeons. An example is the Rock Dove, whose urbanized descendants, the Feral Pigeons, choose ledges and niches of buildings or under bridges as the nearest approach to their ancestral cliffs that they can find. High in the Himalayas and neighboring ranges, Snow Pigeons raise their families in caves and crevices of the rugged mountains. Like other cliff-nesters, including certain swallows, parrots, and other pigeons, Snow Pigeons often breed in colonies because suitable sites are concentrated rather than well distributed. Other pigeons that nest in caves or among rocks are Reinwardt's Long-tailed Pigeon of eastern Indonesia, the Bare-faced Ground-Dove of the high central Andes, the Gray-headed Zone-tailed Pigeon of the mountain forests of the island of Sulawesi (Celebes), and frequently the Galápagos Dove. Where sites among rocks are not available, Rock Doves and other cliff-nesters may occupy enclosed holes or open cavities in trees.

Nesting on the ground, in scrapes that are often scantily lined, is usual for pigeons of the Australian genus *Petrophassa,* including the White-bellied Plumed Pigeon, the Red Plumed Pigeon, the Partridge Bronzewing,

Wonga Pigeon *Leucosarcia melanoleuca*
Sexes similar. Eastern Australia.

the Bare-eyed Partridge Bronzewing, and the White-quilled Rock Pigeon, which last also lays its eggs amid a few twigs on exposed rock surfaces. In the same region, Flock Pigeons regularly nest on the ground, in a slight depression in the shelter of a low bush or clump of grass, and the Brush Bronzewing does so frequently. In the New World, ground nesting is occasional in the Mourning Dove, the Zenaida Dove, the Common Ground-Dove, the Plain-breasted Ground-Dove, the Croaking Ground-Dove, the White-bellied Dove, and the Key West Quail-Dove.

In New Guinea, the Thick-billed Ground-Pigeon builds, in the alcove between two plank buttresses of a tree, a platform of twigs, on which it fashions a shallow nest of rootlets, a few leaves, and tufts of moss. Ground-nesting is closely but not invariably associated with terrestrial foraging. Many species that find their food mainly or almost exclusively on the ground regularly nest above it, in bushes and trees, as does the terrestrial Wonga Pigeon of Australia. On the other hand, pigeons that forage chiefly in trees and shrubs only rarely nest on the ground, probably when they can find no better site.

The few pigeons that nest in colonies generally do so in protected situations. Rock Doves and Snow Pigeons make their homes in the inaccessible walls of caves and cliffs. Others live on small islets devoid (at least originally) of terrestrial predators; they include White-crowned Pigeons in the Bahama Islands, Nicobar Pigeons in the Indo-Australian region, and Australian Pied Imperial Pigeons. In the lower valley of the Rio Grande in

Texas, White-winged Doves sometimes crowd from two hundred to four hundred or more nests into an acre (five hundred to one thousand in a hectare) of dense woodland, each pair occupying a single tree or part of one; but in arid Arizona and Sonora, White-wings' nests are much more widely scattered among cacti and thorny bushes.

Of all extant pigeons, Eared Doves in Argentina probably form the greatest breeding aggregations. In grain-growing districts of the province of Córdoba, their crowded colonies contain millions of nests spread over a square mile (2.6 km^2) or more of thorny scrub, at a maximum density of about twelve hundred nests per acre (three thousand per hectare). In regions of Argentina without concentrated sources of food, the nests of these same doves are dispersed, although colonial nesting occurs among terrestrial bromeliads (plants of the pineapple family) in Brazil. Probably no pigeon ever bred in vaster, more crowded colonies than those of the extinct Passenger Pigeon in the broad-leaved forests of the eastern United States. Lacking the protection afforded by forbidding cliffs or seagirt islets, colonies of Eared Doves and Passenger Pigeons are, or were, widely known and readily accessible to predators of all sorts. Their viability depends on predator satiation. With the availability of enormous numbers of eggs and nestlings, predatory animals can fill their maws without destroying a disastrous proportion of the population. Losses to predation might be no greater than they would be if the same number of nests were scattered over a wide area where they might be exposed to a larger number of predators. However, in the citrus groves of Texas, researchers found that the success of White-winged Doves with dispersed nests was greater than that of doves in colonies.

Pigeons' nests are made of twigs, straws, tendrils, lengths of vines, and dry inflorescences of trees and shrubs, with occasional green or dry leaves or green moss, as in mossy montane forests. The inner layer, or lining, is often of finer pieces than the foundation. Pigeons that forage on the ground walk over it, picking up bits of plants, shaking them to test their suitability, dropping them if they are too decayed or fragile, and walking or flying to the nest site with them if they are adequate. Thin wire appears to have just the pliability that some pigeons prefer. Where discarded wire is available in suitable lengths, as on vacant urban lots, Feral Pigeons may build their nests partly or wholly with it, and Laughing Doves sometimes use much of this material. Pigeons that find most or all of their food in trees gather their building materials in them, often struggling vigorously to break off a twig or dry inflorescence. After detaching a branchlet from a high treetop, a Scaled Pigeon may carry it several hundred yards to a nest site in a low thicket or overgrown grain field in an adjoining clearing, although this species sometimes nests high in palms or other trees. Pied Imperial Pigeons prefer living twigs with green leaves for their large, strongly

Spotted Dove *Streptopelia chinensis*
Sexes alike. India, Sri Lanka, southeastern Asia, Indonesia. Introduced in California, Hawaii, and Australia.

built nests. Wood Pigeons, which forage much on the ground as well as in trees, may collect materials in either situation.

Unlike many passerine birds, which can fill their bills with several items before they fly to the nests they are building, pigeons carry only a single piece. When they appear to hold more, I believe it is always because the twig or inflorescence is branched. One may ask why pigeons do not save time and energy by collecting several pieces together. Those that pull to break a twig from a tree could not engage in this strenuous work with another piece in the mouth. This consideration could hardly apply to pigeons that gather loose pieces from the ground. Might it not be that they are bound by a limitation inherited from more arboreal ancestors? In any case, it is less necessary for pigeons, which feed their young by regurgitation, to be able to carry several items simultaneously than it is for passerines, which carry whole billfuls of fruits or insects to their nests—and can sing with their mouths full.

The statement sometimes made that the female pigeon builds the nest can be misleading. The male chooses the nest site, often starts a bout of building, and works harder than she does, for he gathers most or all of the materials and carries them to the growing nest, where the female waits to receive them. Arriving with a contribution, he stands beside her or, frequently, on her back, facing the same way she faces, and bends down to deposit the piece beside or in front of her. Sometimes, he has difficulty keeping his footing on the sloping sides of her smooth back and slides off. After he flies away, she pushes the piece into the growing structure. At intervals, she turns to face another direction and add the material to a different side, thereby rounding the nest. Continually billing the pieces and shuffling them together, she works them into a reasonably coherent fabric.

This widespread pattern of building is not invariable. A male Ruddy Ground-Dove or Blue Ground-Dove may begin a spell of building by carrying a piece of material to the nest and sitting there to arrange it. He may repeat this routine before his partner arrives to replace him and remain while he delivers additional pieces to her. I have seen both members of a pair of Blue Ground-Doves sit side by side on the nest at the beginning of a session of building, caressing each other or moving about together, arranging its components. One male Ruddy Ground-Dove had started nests in two neighboring sites. Whichever one he went to, his mate joined him there, showing clearly that she followed his lead. As a pigeon approaches its mate on the nest, it often rapidly quivers its closed wings, in what appears to be a prudent greeting, not likely to attract hostile eyes, as a more elaborate ceremony might.

Occasionally, the roles of the sexes in building are reversed. A male Wood Pigeon may take a large share, sometimes the greater share, of shaping the nest with materials his partner brings to him. For about twenty minutes one morning, Richard Johnston watched a pair of Inca Doves alternately bringing materials to the nest site. The male brought about twice as many pieces as his mate did and, at least during this interval, did not arrange them. Pigeons sometimes add to their nests during the incubation period. Once I saw a female Blue Ground-Dove carry six twiglets to a nest where her consort covered two eggs. Obstructing foliage prevented a clear view of this activity, but I was certain that he did not always receive the twiglets, for on several occasions he left the nest as she approached it, to return as soon as she departed. James and Beth Wiley saw both sexes of White-crowned and Plain pigeons bring materials to their nests during incubation.

Pigeons build their nests mainly before midmorning. A male Inca Dove brought twelve contributions to his nest in as many minutes. A White-winged Dove added seventy-seven sticks to his nest in three hours. The most rapid building I have watched was by a Ruddy Ground-Dove, who brought eleven pieces in sixteen minutes. Rock Doves are sometimes more energetic; some watched by Johnston continued for a quarter-hour to make five or six trips per minute to their nests. Early in the season, a pair of Inca Doves took nine days to complete their nest, but about three days is their usual time. A pair of Ruddy Ground-Doves did nearly all of their building between seven and nine o'clock on two mornings. Another pair built for three or four days. Blue Ground-Doves need no more than three days to finish their frail structures. A pair of White-tipped Doves completed a nest in three or four days. Fourteen trips with material in thirty-three minutes was the most rapid building I saw at this nest. A nest of the related Gray-chested Dove was built in no more than three days.

Some pigeons build their nests with great haste. White-winged Doves, arriving at their breeding grounds in Arizona already mated and with eggs ready to be laid, have been known to make skimpy nests, and to deposit in each a single egg, only thirty-six hours after they first appeared. To finish a more substantial nest and lay the first egg in it, they need no more than three days.

Pigeons' nests are often loosely made. That of the Pink Pigeon of the island of Mauritius is so thin that cuckoo-shrikes can reach and pierce the eggs through the bottom while a parent sits on them, as they were twice seen to do. More substantial nests are built by White-tipped Doves. An exceptionally solid example was a thick, shallowly concave platform about six inches (15 cm) in diameter, not including projecting ends of sticks, and about three and a half inches (9 cm) high. Among the 350 pieces (not including the smallest fragments) that it contained were weed stems, straws, sticks, dry pieces of vines, fragments of fronds of bracken fern (*Pteridium*), rootlets, and the like. The longest stiff pieces were two crooked straws about one foot (30 cm) long; but a very thin, curved piece of vine measured, when straightened, twenty inches (50 cm), and two others were nearly as long. This nest weighed about two and a half ounces (68 g). A much slighter nest of a pair of Gray-chested Doves, only twenty-five feet (7.5 m) away, consisted of 143 pieces and weighed a little over one ounce (30 g). Nests of the Superb, or Purple-crowned, Fruit Dove of Australia's northeastern rain forests contained an average of fifty-four sticks three to six inches (8–15 cm) long, but most had from one to five forks, which, locked together, made the apparently fragile structures strong and difficult to pull apart.

Soon after a nest is completed, laying begins. Sets of two eggs are most frequent for pigeons, but many species lay only one. About half of the species of *Columba* appear regularly to incubate single eggs. The clutch size of some species varies geographically. In Costa Rica, Scaled Pigeons lay one egg; but in Trinidad, at the same latitude, they lay two. As far as known, all fruit doves lay only one egg, as do also the larger imperial pigeons and the big crowned pigeons. Among the scattering of other pigeons throughout the tropics which incubate solitary eggs are five species of cuckoo-doves, the Pheasant-Pigeon, and the Nicobar Pigeon.

Although three or four eggs are occasionally found in a nest, it is doubtful that so many are ever produced in one sequence by a single free female.

Pigeons who, because of some accident or irregularity, lack a nest when they have an egg ready to be laid will deposit it in the nest of another pair of their own species or that of some other kind of bird. In dense colonies of Eared and White-winged doves, three eggs are much more frequent than they are in dispersed nests. When a single egg is found in the nest of a species that normally lays two, its companion has probably been lost. However, in carefully watched nests, only one egg was laid by a White-tipped Dove who earlier in the year had produced two, the usual number.

The eggs of birds that nest in dimly lighted holes and burrows are usually white, but those in open nests tend to be mottled, tinted, or otherwise pigmented, making them less conspicuous. Nevertheless, most pigeons' eggs are plain white. The reason for this appears to be that the parents keep them almost continuously covered by sitting on them from the moment the first is laid, thereby reducing the need for concealing coloration. Moreover, pigeons do not readily abandon their nests when a potential predator comes into view but tend to sit firm, sizing up the situation, not leaving so long as they appear to have a fair chance of escaping detection. When they burst from a nest the enemy has found, protective coloration would be of no avail. Among the few pigeons whose eggs are more or less strongly tinted with buff is the Ruddy Quail-Dove, which, as I have seen at several nests, leaves its first egg exposed much of the time before the second is laid.

Would fewer pigeons' eggs be lost to predators if they were pigmented instead of gleaming white? To answer this question, David Westmoreland and Louis Best painted Mourning Doves' eggs with brown to make them resemble the naturally speckled or blotched eggs of many birds with open nests. As controls for their experiment, they left an approximately equal number of eggs plain white. At three-day intervals, they drove away the incubating parents from one lot of painted eggs and one lot of white eggs, which in each case were left exposed until the parents returned after an absence that probably rarely exceeded three hours. At other nests, the doves were permitted to incubate their painted or white eggs without human interference.

At the nests where incubation was systematically interrupted, the painted eggs survived about twice as well as the plain white eggs. At nests where the experimenters did not interrupt incubation, the painted eggs also survived better, but the difference between them and the plain white eggs was much less, and not statistically significant. This experiment demonstrates that when humans drive pigeons away from their nests but do not take the eggs, a concealing coloration substantially diminishes the loss of eggs exposed to predators in the interval before the parents return. However, when doves are driven away from their nests by hungry predators instead of experimenting ornithologists, the value of such coloration will be much less, because the predators that find the nests will doubtless eat the eggs, whatever their color.

In an often-quoted study, C. O. Whitman reported that captive Mourning Doves laid their first egg between four and six o'clock in the afternoon, skipped a day, then laid the second egg between half-past six and nine o'clock in the morning of the second day. Others have reported greater variability in the times that Mourning Doves lay their eggs. Harold Hanson and Charles Kossack found that captive doves usually laid their two eggs on alternate days, but slightly more than a third of the clutches were laid on consecutive days. One pair kept under observation for three years laid the eggs usually on alternate days but sometimes on consecutive days. The interval between the deposition of the first and second egg is not constant for either the species or the individual; but, as with other pigeons, it tends to be considerably more than twenty-four hours.

Similar inconstancy in the interval between the laying of the first egg and the second has also been reported of captive Laughing Doves. As the season advances and days become longer, Collared Doves tend to lay the first egg later in the afternoon, the second earlier in the morning. On cloudy days, first eggs are laid about thirty minutes earlier in the afternoon and second eggs about forty-five minutes later in the morning.

To learn just when free pigeons deposit their eggs is more difficult because one hesitates to risk interrupting their laying by putting them off a nest to see what it contains. A female White-tipped Dove settles on her newly made, eggless nest in late afternoon or evening, and by early the next morning she has laid her first egg, which could have been deposited at any time during this interval. The single egg is covered by the parents through the first day and the following night. The second egg appears during the following forenoon, sometimes before midmorning, more than twenty-four hours but considerably less than forty-eight hours after the first. At two nests of Blue Ground-Doves, the first egg was laid before half-past four in the afternoon, a day was skipped, and the second was laid during the morning of the second day. Like the White-tipped Dove, the Ruddy Ground-Dove lays her first egg during her evening and night session on the nest, where it is present at dawn. At five nests, the second egg was deposited on the day after the first was seen, but occasionally two or even three days intervened between the laying of the first and second eggs. The White-winged Dove usually lays her first egg late in the afternoon of the day her nest is finished, the second in the morning of the second following day, about thirty-six hours after the first.

Although the scanty available information suggests that pigeons' eggs are usually laid with an interval of more than one but substantially less than two full days, exceptions occur. The Pink Pigeon's two eggs are deposited with an interval of forty-eight hours, at about four o'clock in the afternoon. Other birds with two-egg sets, including hummingbirds and manakins, also lay them about forty-eight hours apart; but tanagers and many other songbirds allow only one full day, or a little more, to elapse between the deposition of their first and second eggs.

In humans and certain other mammals, ovulation is spontaneous, regulated by an internal rhythm. Most birds require some appropriate stimulus, usually a nest ready to receive their eggs. L. H. Matthews demonstrated that a female pigeon readily lays eggs if placed in a cage with a male of her kind, less readily if given a female companion, and not at all if kept alone. However, if she is provided with a mirror in which she can see herself, she will lay! One can prevent Domestic Pigeons from laying by placing in their newly finished nest two alien eggs, which they will incubate.

Plate 1. Scaled Pigeon *Columba speciosa,* male
Sexes similar, female duller and paler.
Southeastern Mexico to western Ecuador and east of Andes through most of South America to northern Argentina.

Plate 2. Band-tailed Pigeon *Columba fasciata*
Sexes similar, female duller. Southeastern Alaska and British Columbia through western United States, Mexico, and Central America, and northern and western South America to Venezuela and northwestern Argentina; Trinidad.

Plate 3. Barred Cuckoo-Dove *Macropygia unchall,* male
Mountain forests from Kashmir through southeastern Asia and Indonesia to Lombok.

Plate 4. Green-spotted Wood-Dove *Turtur chalcospilos*
Sexes alike. Eastern and southern Africa to Cape of Good Hope.

Plate 5. Emerald Dove *Chalcophaps indica,* male
Wooded regions from northern India through southeastern Asia to Indonesia, Philippines, and eastern Australia.

Plate 6. White-bellied Plumed Pigeon *Petrophassa plumifera*
Sexes alike. Arid central Australia and interior of northern tropical Australia.

Plate 7. Passenger Pigeon *Ectopistes migratorius,* male (above) and female (below)
Woodlands of eastern and central North America from southern Canada to Florida. Extinct.

Plate 8. Galápagos Dove *Zenaida galapagoensis*
Sexes similar, female paler. Scrublands of Galápagos Islands.

Plate 9. Blue Ground-Dove *Claravis pretiosa,* male (above) and female (below)
Open woodland and clearings from southeastern Mexico through Central America
to northwestern Peru and east of Andes over most of South America
to northern Argentina; Trinidad.

Plate 10. Sapphire Quail-Dove *Geotrygon saphirina*
Sexes alike. Forests from Colombia to northwestern Ecuador and east of Andes in Ecuador and Peru.

Plate 11. Blue-headed Quail-Dove *Starnoenas cyanocephala*
Sexes similar. Cuba, in undergrowth of lowland forests.

Plate 12. Nicobar Pigeon *Caloenas nicobarica*
Sexes similar. Small wooded islands near larger land masses from Nicobar Islands, eastern Indian Ocean, through Indonesia to Solomon Islands and Philippines.

Plate 13. Luzon Bleeding-Heart *Gallicolumba luzonica*
Sexes alike. Forests of Luzon and Polillo, Philippines.

Plate 14. Pheasant-Pigeon *Otidiphaps nobilis*
Sexes alike. Mountain and foothill forests of
New Guinea, Aru Islands, and neighboring small islands.

Plate 15. Victoria Crowned Pigeon *Goura victoria*
Sexes alike. Forests of northern New Guinea.

Plate 16. Lesser Brown Fruit Dove *Phapitreron leucotis*
Sexes alike. Woodlands of Philippines.

Plate 17. Pink-necked Green Pigeon *Treron vernans,* male
Forests from southeastern Asia, Sumatra, and Java
to Sulawesi (Celebes) and Philippines.

Plate 18. Jambu Fruit Dove *Ptilinopus jambu,* male (above) and female (below)
Forests and mangrove swamps of Malay Peninsula, Sumatra, Borneo,
and neighboring small islands.

Plate 19. Many-colored Fruit Dove *Ptilinopus perousii,* male
Forests, parks, and gardens of Samoa, Fiji Islands, and
Tonga, western Pacific Ocean.

Plate 20. Orange Dove *Ptilinopus victor,* female (left) and male (right)
Forests of Fiji Islands.

Plate 21. Cloven-feathered Dove *Drepanoptila holosericea,* male
New Caledonia and Isle of Pines, southwestern Pacific Ocean.

Plate 22. Seychelles Blue Pigeon *Alectroenas pulcherrima*
Sexes alike. Seychelles, western Indian Ocean.

Plate 23. Green Imperial Pigeon *Ducula aenea*
Sexes alike. Woods and mangrove swamps from India and Sri Lanka through southeastern Asia and Indonesia to Philippines.

Plate 24. Topknot Pigeon *Lopholaimus antarcticus*
Sexes alike. Rain forests and eucalyptus forests of eastern Australia from Cape York Peninsula to Victoria.

6 Incubation

As we have seen, pigeons sit on the nest for long intervals while it is being built, and they continue to do so after it is finished. Female Ruddy Ground-Doves sleep on the nest before it contains an egg. When the nest has only the first egg, a pair of these doves keep it almost constantly covered during the day by replacing each other two or three times, a schedule that is quite different from the one they will follow after the second egg appears. Then, as with all pigeons, from the smallest doves to the big crowned pigeons, with a single known exception among free birds, the male takes one long session each day and the female covers the nest all the rest of the time. However, in aviaries where mated pairs are closely confined, they may replace each other on the nest more frequently.

The hour of the male's arrival to relieve his mate, who has continued to sit after incubating the eggs through the night, varies from species to species, from individual to individual, and from day to day at the same nest (Table 1). Usually, he arrives about the middle of the forenoon, with, in

Table 1. Incubation schedules of pigeons

Species	Morning changeover[1]	Afternoon changeover[1]	Length of male's session	Locality
Wood Pigeon	ca. 10:00	ca. 5:00	ca. 7 h	England
Band-tailed Pigeon	10:00–10:30	4:30–5:30	6–7½ h	California, United States
Band-tailed Pigeon	8:15–8:30	3:30–4:00	7–8 h	Guatemala
Pink Pigeon	ca. 11:00	ca. 5:00	ca. 6 h	Mauritius
White-winged Dove	8:00–9:30	3:30–4:00	6–8 h	Southwestern United States
White-winged Dove	8:32	5:15	8 h, 43 min	Guatemala
Ruddy Ground-Dove	8:15–8:45	3:15–5:00	7–8¾ h	Costa Rica
Ruddy Ground-Dove	12:43	5:41	4 h, 58 min	Guatemala
Ruddy Ground-Dove	10:00–11:00	3:00–4:00	5–6 h	Suriname
Blue Ground-Dove 1	8:30–9:40	1:30–3:00	4¼–6¼ h	Costa Rica
Blue Ground-Dove 2	7:30–8:30	1:30–4:00	5½–8 h	Costa Rica
White-tipped Dove	9:00–12:00	3:00–4:00	4–6 h	Costa Rica
Ruddy Quail-Dove	6:30–8:10	3:00–5:00	8–9 h	Costa Rica
Superb Fruit Dove	ca. 8:00	ca. 5:00	ca. 9 h	Australia
Australian Pied Imperial Pigeon	none	ca. 4:40	ca. 24 h	Australia

1. When only one time is given for each changeover, the record is for a single day.

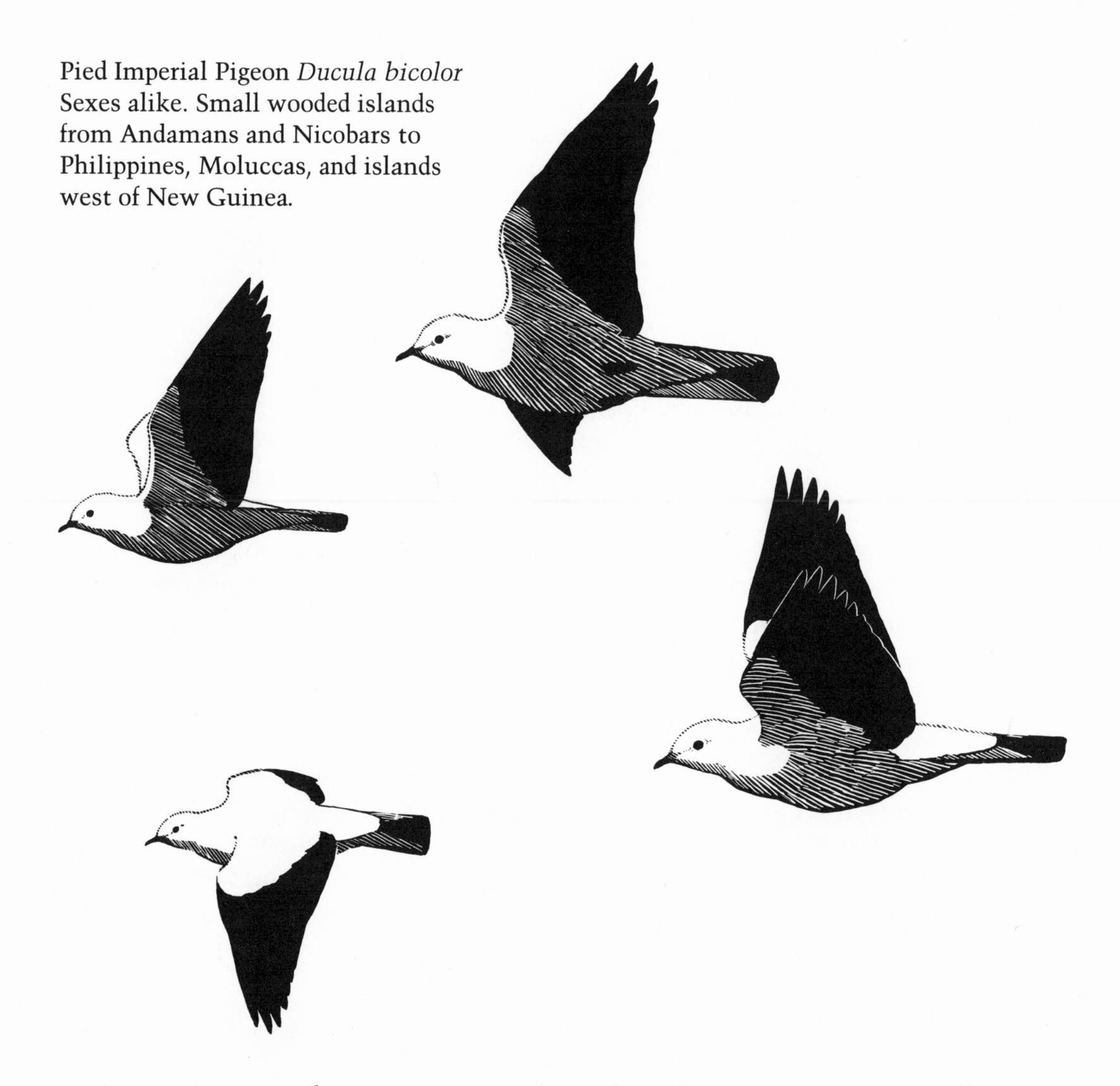
Pied Imperial Pigeon *Ducula bicolor*
Sexes alike. Small wooded islands from Andamans and Nicobars to Philippines, Moluccas, and islands west of New Guinea.

some species, a tendency to come earlier when the eggs are newly laid and when they are about to hatch. Rarely, he delays his arrival until midday. Thenceforth he is continuously in charge of the nest until his partner replaces him in the afternoon, to remain until he returns the following morning. The male's daily session may be as short as four hours or as long as nine. A female who apparently has had difficulty foraging on an afternoon of tropical downpours may relieve her mate so late that he has little time to find his supper after an unusually prolonged turn on the eggs.

Pigeons rarely interrupt their sessions for more than a few minutes, perhaps to avoid soiling the nest, unless they are disturbed, or, especially in hot, dry weather, to drink. Ordinarily, they do not neglect their eggs long enough to forage. The pigeons' pattern of incubation, with one long session by the male through the middle of the day, is similar to that of trogons, at least the New World species, except for the Resplendent Quetzal.

The one known exception to the pigeons' typical schedule of incubation is that of the Australian Pied Imperial Pigeons. These large pigeons nest colonially on small islands, from which they fly over the water to seek fruits in mainland forests. Because the quest for food takes them so far from their nests, they remain away longer than other pigeons. Each partner incubates for about twenty-four hours and is replaced by the other in the late afternoon. Thus, these pigeons have only one changeover each day instead of the usual two. Probably, the same will be found true of other pigeons that nest on islets and forage far away.

Feathers are poor conductors of heat. To warm their eggs more efficiently, birds of many kinds shed all the down from one or more areas on the abdomen, in each of which the skin thickens and becomes well supplied with blood vessels. These bare incubation patches, also called brood patches, are applied directly to the eggs by the incubating parent. As a rule, only the birds (of either sex) who incubate develop these heated pads. Mourning Doves, and probably other pigeons, lack typical incubation patches. During all seasons, adults of both sexes have on the abdomen an area of bare skin, the ventral apterium, which hardly changes when the eggs are laid. Lacking a rich supply of warm blood, it is less efficient than the incubation patch as a source of heat, but it is adequate for the pigeons' needs.

An edematous incubation patch rapidly rewarms eggs that cool when left exposed, as happens repeatedly when one incubating parent alternates sessions and recesses throughout the day. Unless disturbed, pigeons rarely leave their eggs uncovered long enough for them to lose much heat, so rapid rewarming is unnecessary. The Superb Fruit Dove holds its single egg off its thin nest between its thighs, in contact with the bare abdominal patch, and completely hidden by the feathers of its belly. As the bird leaves the nest, the egg drops into it. This method of incubation is hardly feasible for pigeons with two eggs.

A pigeon arriving to replace its mate on the nest does not alight on or close beside it, as many a small bird does. Often, it comes to rest in another part of the supporting tree or branch, where it may delay for a short or long while, preening, before it advances to a point from which it can walk to the nest from a foot to several yards (30 cm to several meters) away. As its partner approaches, the incubating bird rises and walks out along a branch before it takes wing, sometimes with noisy flapping. Such departure avoids the risk of shaking the eggs from a shallow nest by taking off directly from it. After its partner leaves, the new arrival goes to the nest, perhaps murmurs softly as it looks down at the eggs, adjusts them beneath itself, and settles down for a long spell of incubation. Usually the eggs are exposed only briefly at the changeover. Unlike more arboreal pigeons, the Ruddy Quail-Dove approaches by walking over the dimly lighted forest floor, frequently bearing a small leaf or petiole to add to its low nest, to which it flies when it has come near.

I have watched many changeovers during incubation at nests of seven species of tropical pigeons without witnessing a greeting or ceremony more elaborate than wing-quivering, and I have rarely heard a sound other than a subdued coo or that made by the birds' wings on the inward or outward flights. However, Wood Pigeons appear to be different from those I have studied, including the related Band-tailed Pigeon. Murton records that, at the changeover, Wood Pigeons give various appeasement displays, including the submissive posture with lowered head, accompanied by wing-quivering as in nest-calling, sometimes while the birds utter a soft *cooo.* Occasionally the partners caress as they replace each other on the eggs. Even more elaborate were the changeovers of the White-crowned Pigeons watched by James and Beth Wiley in Puerto Rico. Often, one partner remained on the nest while the other stepped up to it. The incoming bird frequently cooed or growled; the outgoing partner gave the nest call. They nodded to each other, or with soft nest calls they preened each other. More rarely, the new arrival fed its consort. In Kenya, V. G. L. Van Someren noticed that the male Red-eyed Dove frequently feeds his incubating partner.

Incubating pigeons sit immobile for long intervals. Occasionally, they drowse, closing their eyes for no more than a few seconds. From time to time, they turn their eggs or preen. Wood Pigeons molt while they incubate, leaving tufts of detached feathers on their nests. As told in the preceding chapter, while one partner sits, the other may add material to the nest. Toward the end of a long spell of incubation, and especially if relief is delayed well beyond the usual hour, the incubating pigeon may become restless, call softly, shift its position, stretch its wings, preen, perhaps fly away; but rarely does it remain absent for many minutes unless it is sure that its partner has come to take charge. Pigeons are reluctant to leave their eggs exposed. Sometimes the incubating bird delays so long in relinquishing the nest to the partner who has come to relieve it that it must be pushed off.

Where nests are close together, male Wood Pigeons, White-crowned Pigeons, and White-winged Doves sometimes interrupt incubation to chase away other males of their species who invade their territory, or to fight with them if they resist expulsion. Females rarely leave their eggs exposed to confront intruders but instead remain on the nest, watching them intently.

If one comes suddenly on an incubating pigeon, it may fly directly from the nest so abruptly that it rolls out an egg or two. Such casual encounters create the impression that pigeons are flighty birds that jeopardize their eggs by panicky departures. Prolonged watching from concealment corrects this view. In long vigils before a nest of Band-tailed Pigeons in a pine tree

White-crowned Pigeon *Columba leucocephala*
Sexes alike. Southern Florida, Bahamas,
Greater and Lesser Antilles south to Antigua, and
small islands of western Caribbean Sea.

high on a Guatemalan mountainside, I recorded their reactions to various situations. Most of the time, when perfectly at ease, each member of the pair kept its head retracted between its shoulders, turned to the left. Distant noises, except when very loud and sharp, such as the sound of blasting, were usually disregarded by the sitting pigeon. Sounds from a nearer source made it stretch up its head and peer around. If the noise became more alarming, the bird rose in its nest and prepared to flee. When an approaching animal proved to be a horse or a cow, snorting or snapping dry twigs underfoot, the pigeon settled down again, then very slowly retracted its head and turned it leftward. Three times during my watches, men seeking cattle or firewood passed beneath the nest without frightening away the incubating pigeon, who lifted its head, saw that it had not been detected, and decided to risk staying with its egg.

On the same mountain, I passed a whole day in a blind before a White-winged Doves' nest in a viburnum tree beside a rivulet at the foot of a pasture. This pair had many visitors. The incubating female remained indifferent to a Steller's Jay, who, early in the morning, collected nest material from the ground nearby. Toward noon, a twittering pair of Bushtits gathered downy feathers, probably shed by the doves, among the branches below the nest, coming within a foot (30 cm) of the male, who appeared not to notice these tiny, bustling birds. When two horses waded up the stream beneath the nest, the dove merely raised his head to learn the source of the sound he heard. He was equally unperturbed when a bull and three cows came running down the slope toward him, then drank and waded in the stream and cropped the lusher herbage on its banks, sometimes directly beneath him. Yet I could not, with the utmost caution, approach within twenty-five feet (7.6 m) of the nest without sending off the sitting bird.

Like the Band-tailed Pigeons, these White-winged Doves were alert to sounds, looked about to discover their source, assessed the situation, and stuck to their post so long as this seemed prudent. They did not risk betraying its location by a premature or unnecessary departure. When pigeons burst precipitately from a nest, perhaps knocking out their eggs or nestlings, it is because they have been discovered by a nest-robber, often man, too powerful to be resisted, and their eggs or young could not be saved by remaining. However, as will presently be told, sometimes they valiantly confront the threat. Students of animal behavior regard pigeons as among the most intelligent of animals, readily learning complicated signals and remembering well. The more intimately I have known free pigeons, the more my respect for their intelligence has grown—as has happened in the cases of a number of animals whose mentality is commonly underrated. In general, birds have brains about as large as, and sometimes larger than, those of mammals of comparable size, and they are not inferior in their ability to learn.

Captive pigeons sometimes incubate in unconventional ways. Derek Goodwin had a female Domestic Pigeon who regarded him as her mate. After a casual affair with a racing pigeon, she laid two eggs and undertook to incubate them without a male's help, sitting through the middle of the day, when properly wedded female incubating pigeons normally are free, leaving her nest only long enough to eat and drink. She did, however, persistently try to call Goodwin's attention to his duty as her partner. Whenever he put his hand into her wicker cage and over her eggs, she would greet him with the halting *coo* of a hen pigeon being relieved by her consort and would leave with a contented air. As long as he remained at the cage, she would bathe and loaf; but as soon as she noticed he had deserted his post, she would fly up to investigate and, finding her eggs uncovered, would promptly settle on them. Such behavior is usual with pigeons; in case of need, either sex will incubate eggs or brood nestlings "out of turn," as is well known to pigeon-racers who send one member of a breeding pair to compete in a contest. In twenty-one days, several more than the usual incubation period, Goodwin's steadfast female hatched her two eggs.

After raising one fledgling (its sibling died), this female Domestic Pigeon again started to incubate two eggs fertilized by a male with whom she was not paired. When her surviving offspring, a male, was less than a month old, he picked up a few twigs and presented them to her on the nest in the conventional manner. At the age of one month, this youngster undertook to incubate his mother's eggs while she took a recess, but he had trouble keeping them beneath himself. Each day, at about eleven o'clock in the morning, he replaced her; but after about two hours of incubating, or trying to, he would give up sitting and join his mother, who promptly returned to her nest. Usually, he took another turn in the afternoon but became bored and left after an hour or two.

When two Ringed Turtle-Doves whose sex was still unknown were confined together in a large cage, they promptly constructed an excellent nest, then spent much of each day sitting in the eggless structure. Their behavior revealed that both were males. When given a Mourning Dove's egg, the two sat side by side on it all day but abandoned it at night while they roosted on a perch. Before its transfer from the Mourning Dove's nest to the turtle-doves' nest, this egg had been left unattended for four days; then, during twenty-one days, it had been incubated only in the daytime and left exposed to nocturnal temperatures of sixty-five to seventy degrees Fahrenheit (18–21°C). Nevertheless, it produced an apparently normal hatchling after nearly twice the Mourning Dove's usual incubation period of about two weeks, thereby demonstrating the egg's great resistance to interrupted incubation and chilling. The foster fathers fed the tiny nestling; but by trying to brood it together, they succeeded only in smothering it.

A single male Mourning Dove in an aviary did better than the two male turtle-doves. When he lost his mate only four days after the incubation of their two eggs began, he sat so steadily that both eggs hatched. He alone fed and brooded the nestlings until they fledged, three or four days beyond the usual age. Despite his valiant efforts, the fledglings were underweight, and the lighter of them died three days after leaving the nest.

Birds of many families will continue to incubate their eggs well beyond the normal incubation period, thereby providing a margin of safety for late-hatching chicks. If the eggs prove to be unhatchable, the parents may attend them for two or even three times the usual incubation period. The record for prolonged incubation appears to be held by a pair of Mourning Doves who, in the exceptionally warm November of 1971, built a nest and laid two eggs on the windowsill of an apartment in Indianapolis, Indiana. When these eggs failed to hatch after the usual fourteen or fifteen days, the birds continued to incubate them. Throughout the following winter, they sat with hopeful tenacity, even on days when the thermometer fell far below the freezing point and when snow blanketed the incubating dove. On the whole, however, the weather was mild, and an abundance of food at a neighboring feeder made life easy for these birds. When the last egg disappeared, on April 4, 1972, this pair had incubated for an amazing 133 days, more than eight times the normal incubation period. Even then, the female dove sat restlessly on her eggless nest until shooed away by the woman who wished to clean her window. For the first two months, this pair followed the standard incubation schedule of two changeovers daily. By February they were replacing each other three times a day, and by March the number of changeovers had increased to seven in a full day.

Pigeons' strong attachment to their eggs, even when contact with them is unpleasant, if not painful, was demonstrated experimentally by Edwin Franks, using Ringed Turtle-Doves. He made hollow artificial eggs, with two attached tubes through which water (with alcohol added at subfreezing temperatures) could be circulated at controlled temperatures ranging from 25 to 144 degrees Fahrenheit (−4–62°C). The normal incubation temperature of Ringed Turtle-Doves is 100 to 102 degrees Fahrenheit (38–39°C).

Despite the extreme temperatures in contact with the dove's sensitive abdominal skin, they continued tenaciously to cover the artifacts. On eggs cooled to the freezing point, they sat with necks retracted and feathers erected to increase the insulation of their bodies, as in cold weather. Nevertheless, they shivered, frequently shaking their heads and closing their eyes. On hot eggs, they sat with neck extended, plumage sleeked, and mouth partly open while they tried to dissipate heat by gular flutter. The overheated doves frequently turned to face another direction, tried to shift the artificial eggs with bill or feet, preened with more than normal frequency, and peered around much. The partners replaced each other on the hot but not the cold eggs more often than they normally do. As long as the experiments continued, they incubated the heated or chilled eggs, trying pertinaciously, in a situation without precedent in the history of their race, to save them from extremes of temperature lethal to embryos. Birds who incubate with alternating sessions and recesses throughout the day tend to spend more time on the nest in cool than in warm weather, but these doves' attendance of their eggs was independent of the eggs' temperature.

Although pigeons, like other birds, have excellent color vision, at least some of them incubate their eggs regardless of their color, as Elliott McClure demonstrated by painting the eggs of free Mourning Doves with children's watercolors. The colors applied to them were red, orange, yellow, green, blue, and black. In some tests, one egg was tinted and the other left white; in other trials, both eggs of a set were painted, each with a different color. One egg was decorated with red, white, and blue and its companion in the nest with brown. Finally, both eggs of a set were striped yellow and black with a white band around the middle. If not taken by predators, all these strangely colored eggs were hatched by the parent doves, who promptly returned to them. Probably Mourning Doves are tolerant of

Speckled Pigeon *Columba guinea*
Sexes alike. Tropical Africa; a disjunct race in southern Africa.

changes in the color of their eggs because, after a rain, the shells often become clay-color or black from the mud in which the birds walk while foraging. This coating is rubbed off by the breast feathers of the incubating doves, but the eggs never regain their original whiteness.

With wide experience of unconfined Mourning Doves, McClure concluded that their fidelity to their nests was probably not exceeded by that of any other free bird of North America. Short of destroying their eggs or young, little could be done to discourage the parents from attending them. However, he noticed one exception to this generalization. These doves frequently desert a nest with an egg that is slightly cracked or punctured, although sometimes they carry away the damaged egg and continue to incubate the other. Since an egg with a fractured shell probably would not hatch, the doves do well to abandon it. Doves returning to their nest usually scrutinize their eggs before they settle down to incubate them, but apparently they learn an egg's condition by touch. A dove who settles on a punctured eggs quietly rises to examine it by turning it over with its bill, often repeatedly. If the egg does not feel right, it is pushed aside or carried away.

Pigeons take a day or more to break their way out of the shell. One egg of the Blue Ground-Dove was pipped more than 25 hours, another more than 32 hours, before each hatched. In the Ruddy Quail-Dove, the interval between the appearance of the first tiny star-fracture that revealed the approach of hatching and the nestling's escape from the shell was more than 21 hours in one instance and at least 18 hours in another.

Corresponding to pigeons' great range in size and habits, the length of their incubation periods varies enormously. For most species, the duration of incubation, from the laying of the last egg of a set to the hatching of this egg, is unknown; and a large portion of the periods that we know were determined in aviaries rather than in the wild (Table 2). Most usual are periods in the range of 12 to 14 or 15 days for the more familiar species of temperate and tropical countries. Pigeons of the genus *Columba,* including Domestic and Feral pigeons, take longer to hatch their eggs, usually 16 to 18 days. Very long periods, about four weeks, are recorded for larger species, including crowned and imperial pigeons. The shortest incubation period reported for any pigeon—indeed, one of the shortest incubation periods for all birds—is that of the Ruddy Quail-Dove, which was only 11 days at several nests in predator-ridden Costa Rican rain forests and in a European aviary.

The interval between the hatching of the first and second egg of a two-egg clutch is often hours less than that between the laying of these eggs, which suggests that the parents do not apply much heat to the first egg before the second is laid. Probably, they cover the single egg to conceal its revealing whiteness rather than to incubate it.

Table 2. Incubation and nestling periods of pigeons

Species	Incubation period (days)	Nestling period (days)	Locality
Rock Dove (*Columba livia*)	16–18	ca. 35	Faeroe Islands
Stock Dove (*Columba oenas*)	16	26–28	Great Britain
Speckled Pigeon (*Columba guinea*)	16	25	South Africa
Wood Pigeon (*Columba palumbus*)	16–17	22	Great Britain
White-crowned Pigeon (*Columba leucocephala*)	13–15	17–25 (21.3)[1]	Puerto Rico
Band-tailed Pigeon (*Columba fasciata*)	—	28–29	United States
Turtle-Dove (*Streptopelia turtur*)	13–14	11–12	Europe
Collared Dove (*Streptopelia decaocto*)	14–16	15–17	Europe
Ring-necked Dove (*Streptopelia capicola*)	12	16–17	South Africa
Green-spotted Wood-Dove (*Turtur chalcospilos*)	13	13	Captive
Masked Dove (*Oena capensis*)	14	12	Captive

Table 2. Continued

Species	Incubation period (days)	Nestling period (days)	Locality
Emerald Dove (*Chalcophaps indica*)	14–16	12–13	Captive
Brush Bronzewing (*Phaps elegans*)	16	16–20	Captive
Diamond Dove (*Geopelia cuneata*)	12–13	11–12	Captive
Mourning Dove (*Zenaida macroura*)	14–15	13–15	United States
Eared Dove (*Zenaida auriculata*)	14	12–14	Argentina
Galápagos Dove (*Zenaida galapagoensis*)	13	17	Captive
White-winged Dove (*Zenaida asiatica*)	12–15 (14)[1]	14	Southwestern United States
Common Ground-Dove (*Columbina passerina*)	13–14	11	United States
Ruddy Ground-Dove (*Columbina talpacoti*)	12–13	12–14	Costa Rica, Suriname
Croaking Ground-Dove (*Columbina cruziana*)	14	10–11	Southwestern Ecuador
Inca Dove (*Columbina inca*)	13–14	14–16	Southwestern United States
Blue Ground-Dove (*Claravis pretiosa*)	14–15	11–14	Costa Rica
White-tipped Dove (*Leptotila verreauxi*)	14	15–17	Costa Rica, Argentina
Ruddy Quail-Dove (*Geotrygon montana*)	11	10	Costa Rica, Captive
Nicobar Pigeon (*Caloenas nicobarica*)	—	ca. 3 mo.	Captive
Luzon Bleeding-Heart (*Gallicolumba luzonica*)	17	12	Captive
Pheasant-Pigeon (*Otidiphaps nobilis*)	23–26	—	Captive
Blue Crowned Pigeon (*Goura cristata*)	28–29	30–36	Captive
Victoria Crowned Pigeon (*Goura victoria*)	30	28	Captive
Wedge-tailed Green Pigeon (*Treron sphenura*)	14	12	Captive
Superb Fruit Dove (*Ptilinopus superbus*)	14	7–8	Australia
Lilac-capped Fruit Dove (*Ptilinopus coronulatus*)	18	12	Captive
Knob-billed Fruit Dove (*Ptilinopus insolitus*)	19	14	New Britain, Southwestern Pacific
Australian Pied Imperial Pigeon (*Ducula spilorrhoa*)	ca. 26–28	24–28	Barrier Reef, Australia

1. Average.

7 The Young and Their Care

A newly hatched pigeon is a soft, helpless creature, whose pink or darker skin is sparsely covered with tufts of short, hairlike, straw-colored or whitish down instead of the more plumelike natal down of many passerine birds. Its eyes are tightly closed. The dusky bill becomes blacker toward the end, then abruptly whitish at the tip. At the end of the upper mandible is a hard, light-colored egg-tooth. The lower mandible bears a similar but smaller projection, which apparently helps the larger egg-tooth to break the shell when the chick hatches.

Hatching appears most often to occur before dawn or early in the morning, while the female is still on the nest that she covered through the night. She carries away the empty shell, usually half at a time, unless both parts hang together. If she neglects to remove it, or if an egg hatches while the male is incubating, he removes the shell. Once I watched a male Ruddy Quail-Dove trying to eat the larger part of an empty shell. During the three-quarters of an hour that he was engaged in this difficult endeavor, one of the nestlings he was brooding rose up for food. He fed the hungry hatchling at one side of his mouth while holding the shell in the other side, making me wonder whether the day-old dove had received hard fragments of shell with its milk. Then he dropped the nestling's head and gulped down the remainder of the shell.

As the time for hatching approaches, both parents prepare to nourish their progeny. The walls of the crop, a saclike expansion of the esophagus, thicken until its weight is increased threefold. The crop's epithelial cells become swollen with proteins and globules of fat, then slough off, producing crop milk. About the time the young escape their shells, the parents are ready to feed them with this milk, whose production, like that of mammalian milk, is controlled by the hormone prolactin from the anterior lobe of the pituitary gland. Thick and curdlike, the milk contains 65 to 81 percent water, 13 to 19 percent protein, 7 to 13 percent fat, 1 to 2 percent mineral matter, and vitamins A, B, and B_2. The protein consists of a large variety of amino acids. The mineral component is largely sodium, with a small amount of calcium and phosphorus. Unlike mammalian milk, crop milk is devoid of sugar and other carbohydrates, but it contains the sugar-reducing enzymes amylase and invertase.

Until the nestlings are a few days old, milk is prepared for them only while the parents' corps are empty, thereby ensuring that it is not mixed with more solid food, which the recipients may be unable to digest. As the nestlings grow older, they receive, mixed with the milk, increasing amounts of the food swallowed by their parents, who tend to gather for them more insects and other invertebrates than they include in their own diets. However, Eared Doves in Argentina gave neither fruits nor animal food to their nestlings, only seeds smaller than those commonly eaten by adults. When newly hatched, Wood Pigeons receive only pure crop milk. Thereafter, the proportion of milk in their diet decreases from 92 percent during the first three days to 49 percent between the ages of seven and nine days, 33 percent between ten and fourteen days, and about 20 percent at greater ages. Domestic Pigeons receive some milk until they are about sixteen days old; but Eared Doves, which develop more rapidly, are given significant quantities only until they are nine or ten days old. (Among the few other birds that produce esophageal milk for their young are male Emperor Penguins and Greater Flamingos.) Because of this preparation, a parent pigeon can feed the newly hatched nestling without first going off to hunt food for it, as few other birds can do.

Lacking such specially prepared food, most parent birds, especially passerines, feed their young directly from their bills. Each time a parent comes to the nest, it promptly gives to one or more members of its brood what it has brought for them, then flies away for more or settles down to brood. The number of meals delivered tends to equal the number of parental visits to the nest during intervals of active provisioning.

The pigeon's manner of feeding its young is very different. Arriving to attend nestlings only a few days old, a parent may sit brooding for a long while, sometimes an hour or more, probably clearing its crop of the food it has just eaten and then secreting milk, before it begins to feed. Remaining on the nest, it can give in installments, over a long interval, the food that it elaborates within its own body. Early in the nestling period, the number of installments, or meals, may greatly exceed the number of parental visits to the nest. Through judiciously spaced periods of observation, each continuing for a few hours, one can learn very well how nestlings of most small birds are fed; only by continuous daylong watches, can one obtain adequate information about the feeding of nestling pigeons. And while one sits patiently watching, probably in a stuffy blind—for most pigeons are shy—the bird may perversely turn its tail toward the observer, preventing an adequate view of the transactions.

It is not easy to find either a nest of free pigeons wholly satisfactory for studying the care of the nestlings or a person with the patience and dedication adequate for such an undertaking. An exceptional opportunity arose when Mourning Doves nested on the windowsill of Dorothy Luther's apartment in Indianapolis, Indiana; with remarkable determination, she watched continuously, comfortably seated at her window, from dawn to dusk of the fifteen days that the young remained in the nest. Hers is the most complete record of the parental activities of a pair of free pigeons which has come to my notice. One of the results of her study is the very detailed feeding schedule that, with her permission, I present here in slightly simplified form (Table 3).

When drowsy nestlings are slow to take their meals, the parent may prod them gently or touch their bills to arouse them. Hungry nestlings solicit food from their brooding parent by rising up in front of it and ruffling the feathers of its chest or head. If the parent responds, they insert their bills into its open mouth, sometimes up to their eyes. While very young and sightless, nestlings are commonly fed singly; but after their eyes open, two are often fed simultaneously, one on each side of the adult. With obvious muscular effort, the adult regurgitates an unseen stream of aliment directly into their mouths. When the nestlings grow bigger and the stream apparently becomes more copious, bringing up the food requires more strenuous exertion. As with violent, convulsive movements of neck and head the parent pours food into the two mouths pushed into its own, three heads bob

Table 3. Schedule of feeding for two nestling Mourning Doves

Age (days)[1]	Hours watched	Number of feedings by ♀	♂	Both	Total time of daily feedings (seconds) ♀	♂	Both	Duration of feedings (seconds)	Average length of meals (seconds)
0	14.5	7	7	14	190	1024	1214	3–338	86.7
0–1	15.0	10	12	22	1555	1026	2581	4–357	122.4
1–2	14.5	8	8	16	1433	1071	2504	5–471	156.5
2–3	15.0	5	7	12	833	1769	2602	98–745	209.6
3–4	15.0	3	1	4	920	348	1268	160–506	327.3
4–5	14.5	2	1	3	757	364	1121	297–460	371.2
5–6	14.5	2	1	3	581	336	917	216–365	313.2
6–7	14.5	3	2	5	604	616	1220	134–420	254.6
7–8	14.5	3	2	5	461	373	834	71–240	170.0
8–9	14.5	2	2	4	336	300	636	73–216	159.0
9–10	14.5	3	3	6	296	185	481	28–125	80.1
10–11	12.5	4	4	8	195	175	370	16–90	46.25
11–12	14.5	4	4	8	197	183	380	16–125	47.5
12–13	14.5	5	5	10	265	242	507	18–93	50.7
13–14	15.0	4	5	9	149	238	387	12–69	42.4
15	4.5	0	1	1	0	30	30	30	30.0
TOTAL	222.0	65	65	130	8772	8250	17,022		

Source: Modified from Luther 1979a, courtesy of Dorothy Luther.
1. The nestlings hatched on consecutive days.

jerkily up and down together, while the youngsters flap their wings. At the conclusion of the meal, the parent drops or shakes out the bills of the nestlings, who may rise up for more. Although Mourning Doves, White-tipped Doves, Blue Ground-Doves, Ruddy Ground-Doves, and Ruddy Quail-Doves are usually fed singly while they are sightless and two together after both can see, Goodwin found just the reverse with the pigeons in his aviary. After they were well grown, the young birds poked and thrust too vigorously for a parent to deal with more than one at a time. Male and female pigeons commonly take equal shares in feeding their offspring.

Captive Pheasant-Pigeons behave in ways that appear to be unique in the family. While their single egg is hatching, both parents sit on the nest with heads together and lowered close to it. During the first week after hatching, the father feeds the mother on the nest and she feeds the young. Such indirect feeding of nestlings, by no means rare among other birds, is unusual among pigeons. However, V. G. L. Van Someren watched a male Tambourine Dove regurgitate to his mate while she brooded newly hatched nestlings.

Table 3 shows that the meals of very young Mourning Doves are brief and frequent. The nestlings in this table hatched a day apart. On the first day of observation, when only one was present, the parents fed it 14 times, each meal lasting from 3 to 338 seconds. On the second day, two nestlings were given, separately, a total of 22 meals, doubling the time devoted to feeding. On the third and fourth days, the meals were fewer but longer; the total time occupied by feeding remained almost the same. After this, the number of meals per day decreased sharply until, through most of the nestling period, the two young received, usually simultaneously, from 3 to 8 meals daily. The duration of single meals continued to increase until, for nestlings four and five days old, it averaged slightly more than 6 minutes. When the Mourning Doves were thirteen and fourteen days old, they received 9 meals, totaling 6.5 minutes. Most meals now took less than a minute.

At a nest of the very different Ruddy Quail-Doves, feeding the nestlings followed a similar course. Two day-old nestlings were fed 14 times by their father and 8 times by their mother—22 meals in a day. Each feeding (while a nestling's bill remained continuously in a parent's mouth) lasted from a fraction of a minute to 3 or 4 minutes, rarely 5 or 6. The 14 feedings by the male totaled about 21.5 minutes, the 8 feedings by the female 23.5 minutes. Together, the parents fed for 45 minutes. When the same two nestlings were four days old, they were fed 3 times by each parent, a total of 6 meals. The feedings by the male lasted 9, 4, and 2 minutes; those by the female lasted 15, 8, and 4 minutes. Together, they fed the nestlings for 42 minutes. When these nestlings were seven days old, they were fed once by their father, for 10 minutes, and twice by their mother, each time for 6 minutes. The three meals lasted a total of 22 minutes.

At another Ruddy Quail-Doves' nest, the father held the bills of his two six-day-old nestlings in his mouth continuously for 25 minutes, but in this long interval he regurgitated only intermittently, as was evident from the vigorous movements of his head and body, including alternate expansion and contraction of his chest. These efforts were separated by longer intervals of repose. Such intermittent regurgitation was typical of most of the longer feeding sessions. Since it was not feasible to measure the exact time devoted to regurgitation, each long session was counted as a meal. At still another quail-doves' nest, nestlings ten days old and almost ready to fly

Tambourine Dove *Turtur tympanistria*, male
South-central and southern Africa.

were fed once by their father and twice by their mother, for a total of only 13 minutes in a whole day. As nestling pigeons grow, their meals become fewer but undoubtedly bigger.

At a Blue Ground-Doves' nest, the number of meals was 28 when the two young were newly hatched, 25 at three days of age, and 8 at six days. The total time of feeding on these days was, respectively, 55, 81.5, and 23 minutes. At a different nest, a solitary week-old Blue Ground-Dove received in a day 18 meals, which occupied a total of 23 minutes. A single day-old White-tipped Dove was fed 11 times, for a total of about 21 minutes. When this nestling was a week old, it was fed twice by its father and once by its mother, for a total of 7 minutes. At fifteen days of age, its mother fed it twice and its father 3 times, for a total of less than 5 minutes.

Other young pigeons are fed less frequently. Only the male of a pair of Band-tailed Pigeons in Colorado fed their single nestling until after its twentieth day, when the female began to nourish it. She did not even feed it on its fourth day, when its father failed to appear and she brooded it continuously while it fasted. During its first week, the father fed it 3 times daily, between noon and three o'clock in the afternoon. During the second week, he gave it 2 meals, between noon and half-past one. After the mother ceased to brood, on the twentieth day, she apparently helped to nourish her nestling. At a Band-tails' nest in California, feeding was similar. During the single nestling's first five days, it was fed about 3 times daily. After brooding ceased, on the nineteenth day, it was fed twice and sometimes only once daily, between eleven and two o'clock. Wood Pigeons in England, which usually have two nestlings, also feed them less frequently as they grow older, until, when the nestlings are eight days old, the number of daily meals is reduced to 2 by each parent. During their last week in the nest, White-crowned Pigeons in Puerto Rico were fed more often, from 5 to 11 times per day. Throughout the nestling period, the father was the chief, and on some days the only, provider.

In the first days after they hatch, nestling pigeons are brooded by parents sitting continuously, according to the schedule they followed while they incubated the eggs, with a tendency, in some species already evident before the eggs hatch, for the male to arrive earlier in the morning. After the first six days, a pair of Mourning Doves changed their routine, each parent coming to the nest twice instead of once, resulting in four daily changeovers instead of two. A few days later, the number of changeovers by this pair increased to six; and during the last few days that the young remained in the nest, the parents visited them up to eleven times, in no regular order. After the first week, the young were left alone for increasingly long intervals in the daytime. At one nest, young Mourning Doves were brooded every night until they flew; but at another nest, they slept alone during their last five nights in it. At this nest, diurnal brooding, for short intervals, continued two or three days after nocturnal brooding was discontinued.

A pair of Ruddy Ground-Doves in Costa Rica began to leave their nestlings exposed when they were only two or three days old. Even under a shower, one brood of this age remained uncovered. Sometimes, the parents carelessly stepped on their nearly naked nestlings, apparently without hurting them. When one brood was four days old, each parent came twice to feed the nestlings, making four changeovers in a day instead of two, as during incubation and the nestlings' first days. Blue Ground-Doves three days old were exposed for over two hours. When these nestlings were six days old, the parents changed their schedule, and each visited them twice in a day. Both Ruddy Ground-Doves and Blue Ground-Doves brood their nestlings by night until they fly. Nestling Ruddy Quail-Doves are left unattended long before they are well feathered, even during showers. One week-old brood remained alone for half a day. White-tipped Doves, among the most devoted of avian parents, brood their nestlings, or at least rest on the nest, guarding them, almost continuously throughout the nestling period, even when this is abnormally prolonged by a few days. Only during the final days do they change their routine from two to four changeovers daily.

As they grow older, nestling pigeons often rest with their heads protruding from the brooding parent's breast; on mild days, they may lie exposed in front of the adult, who gently preens their sprouting plumage. At intervals, they preen themselves, perhaps while the parent does the same for itself, and occasionally they preen nest-mates. They flap their wings vigorously, sometimes in the guardian parent's face. One young White-tipped Dove became most active in the brief intervals between the departure of one parent and the arrival of the other, moving around, preening, stretching both wings together, flapping them, as though enjoying the freedom to exercise without a larger body looming above it. When rain falls or night approaches, the youngsters retire beneath their parent; if they have grown too big to be comfortably covered, they push their heads under the sheltering breast, leaving much of their bodies exposed.

Although nesting pigeons normally follow a fixed routine, their parental behavior can be flexible. In the preceding chapter, we saw how a single parent of either sex can undertake the duties of its missing partner and, unaided, rear its young. Pigeons' attendance of unfledged young is less tied to the nest in its original site than is that of some passerines. In Panama, a pair of Ruddy Ground-Doves built a nest on a bunch of green bananas. When this bunch was cut down, without injuring the nestlings, still in pinfeathers, I placed their nest on a neighboring bunch, where the parents continued to attend their brood. Two White-tipped Doves, barely able to fly, were frightened from their nest. I captured and replaced one of them. The

Ruddy Ground-Dove
Columbina talpacoti, male
Central Mexico to
northern Argentina.

other ran and fluttered over the ground too well to be caught. For two or three days, the parent doves attended one of their young on the nest and the other on the ground, until it could rejoin its sibling still on the nest. In Maryland, parent Mourning Doves brooded a nestling about five days old that had fallen to the ground. After its fallen nest was tied in a tree, not in its original site, the parents continued to attend it there until it fledged.

In colonies of Australian Pied Imperial Pigeons on small islets, disturbed young leap from their nests long before they can fly, flutter to the ground, and flap their way to shelter, often in a tangle of roots. When all is quiet, they return to their nests, climbing with feet, wings, and bill, the latter used as a hook—much as a young Hoatzin regains its nest above the margin of a South American river. One callow imperial pigeon covered fifty feet (15 m) on the ground and climbed five feet (1.5 m) to reach its nest. As a flightless youngster passes other nests on its way back to its own, the attendants of these nests chase and attack it, buffeting it with their wings and pecking feathers from its head and back. Such surprising mistreatment of straying chicks is not uncommon among other colonial-nesting birds, including certain terns, gannets, and albatrosses. They may attack the chick to prevent its joining another in a nest where the parents are unable to nourish adequately more than one.

At all times, Ruddy Quail-Doves and White-tipped Doves keep their nests irreproachably clean by swallowing all the nestlings' droppings, often shortly before feeding them. One wonders whether they mix the excrements with the young birds' food and, if not, how they avoid this. Possibly the droppings bypass the crop, where the food for regurgitation is stored. The Mourning Doves on Dorothy Luther's windowsill ate the soft, warm droppings of nestlings only a few days old as soon as they were voided, "as though they were gustatory delicacies." The father stimulated nestlings one and two days old to defecate by running his beak beneath their wings and then pecking the anus, about half the time with positive results. After the nestlings' first week, these Mourning Doves relaxed their attention to sanitation, with the result that the nest and supporting ledge soon became heavily soiled. Although I have been unable to watch Scaled Pigeons and Red-billed Pigeons closely enough to see whether they removed droppings, I have found their nests clean while the nestlings were still unfeathered. After the nestlings were older, the nests became soiled around the edges, although the center remained clean. Probably, like Mourning Doves, these pigeons eat droppings only while the nestlings are young. Week-old White-crowned Pigeons occasionally drop their excreta over their nest's edge.

Blue Ground-Doves, Ruddy Ground-Doves, and Inca Doves apparently give no attention to sanitation. Even before their eggs hatch, their nests may bear a few droppings deposited by the incubating parents. Before the young fly, the nests become laden with waste. This may bring certain advantages, for in dry climates, like those inhabited by Inca Doves, the accumulated excrement cements the frail nests together, making them more stable, less likely to be blown away in a windstorm. Inca Doves may use the same nest repeatedly, with better results in old, weighted structures than in new loosely built ones. Because observers frequently fail to record this important point, we do not know how widespread in the pigeon family nest sanitation is.

While I watched a Gray-chested Doves' nest in the underwood of a Panamanian forest, a party of Rainbow-billed Toucans twice flew heavily and noisily through the trees above and around it. Each time the great-billed birds appeared, the brooding male dove bowed forward, depressed his neck and head, and slightly spread his wings. This posture made him less conspicuous from above, for his light-colored foreparts were turned downward, while his brownish dorsal plumage was presented to the nest-robbing birds in the trees above him. His spread wings screened the pale plumage of his sides. The crouching bird blended well with the brown sticks of the nest and the dead stems of the vine tangle that supported it. After the toucans passed from view, the dove slowly resumed his usual brooding posture.

Such forward bowing by nesting pigeons when they are disturbed or alarmed is widespread. A brooding Ruddy Quail-Dove depressed his foreparts whenever I approached his nest, when a squirrel passed by, and when domestic chickens scratched around it. When four Fiery-billed Aracaris flew overhead, the dove assumed the crouching posture in its most extreme form, pressing his breast against the nest, elevating his posterior parts until his tail stood straight up, and continuing in this attitude until these middle-sized toucans vanished. In the forests of eastern Australia, the Superb Fruit Dove incubates a single egg while resting near the edge of the nest cup with foreparts depressed and rear part elevated. When an incubating male is approached, he turns away from the intruder, so that only the underside of his tail and his white, green-spotted underparts are visible to him, while the bird's reddish purple cap and brick red shoulders are lowered and hidden. If one walks around the nest, the dove rotates to keep his tail toward the observer. The less intensively colored females were not seen to behave in this fashion; if closely approached, they flew away instead of rotating.

When a pair of White-tipped Doves built their nest in a dense tangle of scrambling *Dicranopteris* ferns behind our house, the timid male would fly away if I peered into the fern where he brooded, but his mate permitted me to touch her lightly before she flew. One evening in the twilight, repeated

White-tipped Dove *Leptotila verreauxi*
Sexes alike. Southern Texas to
central Argentina.

thuds, apparently made by the female striking something with a wing, drew my attention to the nest. My flashlight revealed a fairly large snake that the courageous dove was trying to fend off with strong wing blows. Wiry fern stipes broke the force of the blow that I aimed at the serpent, which vanished into the dark tangle. Although this encounter shook the nest, the dove remained steadfastly sitting. Next morning, one of her nestlings had vanished, whether devoured by the snake or lost in the scuffle I could not learn. Three evenings later, more wing-flapping brought me hurrying to the nest in the twilight, but this time I found no snake. Later that night, the dove was sitting peacefully on her nest. By morning, the second nestling had vanished.

The White-tipped Dove who struck with a wing the stick that I raised to its nest (as told in chapter 3) finally dropped to the ground and began to move away, hopping and limping, quivering its wings or loosely flapping them, as though badly injured and unable to walk or fly. The coffee plantation where this occurred had recently been cleaned of weeds and made an excellent stage for the dove's act. For two hundred feet (60 m) it led me onward, keeping just a safe distance ahead, until at the edge of the plantation a tangle of bushes and vines forced it to interrupt the performance. Flying competently over the obstruction, it resumed its broken-wing act on the far side, where I could not readily follow it.

Some of the most spectacular distraction displays I have watched were given by Red-billed Pigeons who nested in a grove of Mexican cypress trees amid pastures in the Costa Rican highlands. Often, I found a parent brooding, or at least resting beside, a well-feathered nestling in a nest above my head. The parent was so steadfast that sometimes I had to shake the tree to make it leave and reveal what the nest contained. Dropping low, the pigeon would flutter across the open field, beating its wings loosely, as though scarcely able to fly, yet skimming over the short grass faster than I could follow. Thus, it would lead me onward for a hundred feet or more, until it reached a bush or low tree, in which it would alight and continue wildly to flap its wings while watching me advance. When I came closer than it deemed safe, it would drop down and again fly low until it reached the next bush that offered a limb for perching and waving its wings as before. At times, the Red-bill would make yet a third fluttery flight before flying off in the normal way, leaving me several hundred yards from the nest where I had disturbed it. Once, after luring me away in this fashion, a parent returned while I was examining its nest ten or fifteen minutes later, perched at the top of a neighboring cypress tree, and flapped its wings loosely, as it had done while leading me away at the beginning.

I witnessed a still more impressive performance one day when a dog followed me unbidden into the cypress grove. At our approach, a Red-bill dropped from its feathered nestling almost to the ground. As the pigeon fell, the dog leapt toward it, making it flee more swiftly than while fluttering over the ground in the usual distraction display. Nevertheless, it flew slowly, only a foot (30 cm) or so above the grass, leading on the dog, who continued to follow with high hope of catching the pigeon, until the two had passed over the boss of the hillside, beyond view.

Such distraction displays are frequently given by parent pigeons and many other birds, including those as big as Ostriches, especially those whose nests are low. They give the impression that the birds are either badly injured or so torn between the conflicting motives of guarding their progeny and preserving their own lives that their behavior becomes completely disorganized. Far from being distraught or having a fit, they regulate their movements with cool judgment, as they must do to stay close enough to the pursuing dog or other flightless animal to lure it onward and away from their nest or chicks yet far enough ahead for safety. The proof of this is that they rarely, if ever, attempt the broken-wing act unless they have a clear stage, where they are in no danger of becoming entangled; and, as in the case of the White-tipped Dove, they deftly avoid obstacles.

If the parent does not succeed in diverting the enemy from the nest, the nestlings may try to defend themselves, at least among the larger pigeons. When a young Scaled Pigeon in pinfeathers is disturbed, it rises in the nest, puffs out its breast, and raises its wings, all of which makes it look bigger and more formidable. In this attitude, it sways upward and backward, then downward and forward, making, with each forward lunge, a clicking or clacking sound with its bill. As long as it feels itself menaced, the nestling continues to perform rhythmically in this fashion. To make the *clack*, it pushes its lower mandible slightly forward, until the tip rests against the downwardly bent end of the upper mandible. The bill is then slightly open. Apparently, the two mandibles are pressed together until the lower one snaps back into its normal position; striking together along their entire length, they emit the sharp sound. The nestling also darts forward to peck an intruding hand with its bill. After its feathers expand, it strikes with its wings, much as adult pigeons do. Taken in hand, it struggles vigorously without ceasing to clack its bill, and at the same time it hisses slightly. Doubtless, all this belligerent display intimidates small animals; yet some nestlings are taken by predators. The Red-billed nestling has a similar display, but it does not clack as loudly as does the Scaled Pigeon. Both nestlings strike and bite intruding fingers, but not hard enough to cause pain. Young Domestic Pigeons defend themselves in much the same way. Nestlings often cling to their nest to avoid removal.

As the time for their departure approaches, young pigeons may venture a short distance from their nest on the supporting branch or ledge, then promptly return. They exercise their wings vigorously. Unless menaced, they remain on or near their nest until they are feathered and fly well, although they may alight clumsily. Early one morning, when I approached a nest where a Ruddy Quail-Dove guarded her two nestlings, the trio "exploded," each flying in a different direction through the forest undergrowth. Simultaneous departures of parent and fledglings are frequent among White-tipped Doves; the adult often drops to the ground and feigns injury to lure the enemy away while the young fly to safety. An eleven-day-old Blue Ground-Dove left its nest spontaneously, following its mother and flying strongly until beyond view.

Superb Fruit Doves who abandoned their nests when they were only seven or eight days old had been handled, which in many birds causes premature departure. The Ruddy Quail-Dove, whose incubation period is the shortest recorded for any pigeon, has also one of the shortest nestling periods reported for undisturbed young. When only ten days old, quail-doves abandon their nests, flying well. If disturbed at the age of eight days, they flutter down from their low nests and walk competently away over the forest floor, as though they had long been in the habit of taking solitary promenades. At the other extreme, Nicobar Pigeons have remained in their nests in an aviary for nearly three months, which exceeds the known nestling periods of other pigeons so greatly that one suspects some irregularity attributable to captivity. Big crowned pigeons in an aviary stayed in their nests a month or more. Long nestling periods of twenty-five to twenty-nine days have been recorded of several species of *Columba.* On the far-northern Faeroe Islands, Rock Doves linger in their sheltered nest holes or caves long after they are fully feathered, often until they are thirty-five days old. The twenty-two-day nestling period of the Wood Pigeon is exceptionally short for its genus. Many pigeons and doves leave their nests spontaneously when they are eleven to sixteen days old (Table 2).

For a short while after fledgling pigeons abandon their nests, family bonds remain strong. When two young are reared, they tend to forage and rest together, sometimes in contact, as they did while growing up in the nest. At night, they roost with their parents. For at least a week, two young White-tipped Doves and their parents slept in an orange tree, arranging themselves in diverse ways. Sometimes, they perched two and two, both members of a couple pressed together. On other nights, three slept in a compact row, the fourth a few inches away. Or only two rested in contact, with one an inch or two from each side of them.

When the parents built another nest, one young White-tipped Dove sat on it for an hour while work was suspended, preening or fiddling with the sticks. All this time, its sibling perched nearby, or briefly joined it on the nest. Late on a rainy afternoon, after an egg was laid in this new nest, the mother and a juvenile sat side by side on it, as though both were incubating it. Here they remained until it was quite dark, while the other juvenile roosted alone in the orange tree. While the mother incubated on other nights, the two young doves slept in close contact higher in the same tree. Their father now roosted at a distance from the nest, as is usual with free pigeons. Recently fledged Inca Doves also roost in a tree, wing to wing with their parents. Two young Ruddy Ground-Doves flew from their heavily soiled nest at the (usual) age of twelve days, then returned to the nest in the evening, to be brooded through the night by their mother. On the following evening, after two days of activity among the trees, they approached their nest, apparently for another night on it, but were deterred by my watching. Their mother slept alone on the nest.

The most thorough study of the care and behavior of fledgling pigeons was made with Mourning Doves in Alabama by Ronald Hitchcock and Ralph Mirarchi. They equipped thirty-five nestlings with radio transmitters and their nest-mates with colored wing tags, then kept track of them after they flew from their nests at fifteen days of age. For the first three nights after fledging, the young doves frequently roosted with both parents, but thereafter this association became rare, to cease entirely when they were twenty-nine days old. After the first three days, fledglings slept with their mother somewhat less often than with their father, probably because she was preparing to lay again. Up to the age of twenty-seven days, siblings roosted together much more frequently than they roosted alone, and a few continued this habit until they were a month old. Starting at the age of three weeks, other juveniles roosted increasingly with a flock rather than with their parents or nest-mates.

These Mourning Doves were fed consistently up to the age of twenty-seven days by at least one parent, who, after sixteen days, was chiefly their father. In two instances when the male parent disappeared, the mother continued to feed her young until they were a month old. At the age of seventeen days, fledglings began to feed themselves, at first inexpertly but gaining proficiency so rapidly that at eighteen days they could support themselves if adequate food were available within fifty yards (45 m) of their birthplace and at twenty-one days if they could find enough within two hundred yards (180 m). White-winged Doves and White-crowned Pigeons are also fed by their parents until they are about one month old. A Victoria Crowned Pigeon in an aviary was first seen to feed itself when it was a little more than thirteen weeks old, but for some time after this it received food from its parents.

The duration of parental feeding by Mourning Doves and other pigeons until their progeny are about one month old compares favorably with that of many passerine birds in the North Temperate Zone. Tropical birds tend to feed their fledglings to a more advanced age. Young birds often continue to beg for food until they find it more profitable to search for their own than to try to persuade increasingly reluctant parents to feed them. Two Mourning Doves, nearly a month old, taking their food simultaneously from a vigorously regurgitating parent, flapped their wings wildly and pushed their elder around so boisterously that they appeared to be attacking it. Such rough treatment by hungry siblings may be a factor in terminating parental feeding.

Parent doves that hatch a subsequent brood become antagonistic to the progeny with whom a short while before they had been so intimately associated. The young male Domestic Pigeon who took twigs to the nest where his mother attended a subsequent clutch of eggs, and later did his best to help her incubate them (chapter 6), was firmly repulsed by her after these eggs hatched. When reunited with her after an absence of five weeks, the young male recognized her instantly and became intensely excited, flying to her and displaying eagerly, as a male pigeon usually does when he rejoins his mate after several days of separation.

Probably because parents frequently reject their young of one brood when a later brood hatches, pigeons are not known to practice cooperative breeding, widespread among birds permanently resident in mild climates, in which older offspring help parents to rear younger siblings. Moreover, the tight schedule of incubation and the nestlings' special diet of crop milk, normally produced only by parents who have incubated, appears to preclude effective helping by nonbreeding yearlings. However, pigeons occasionally serve as helpers in both intraspecific and interspecific contexts. Mourning Doves, which sometimes adopt and feed fledglings not their own, exemplify the former. In large colonies of White-wined Doves in southern Texas, adults sometimes feed nestlings other than their own offspring. After feeding its two nestlings, a parent Wood Pigeon flew to feed a strange juvenile who begged for food ten yards (9 m) away. At three nests with small nestlings, an intruding juvenile was fed. On four consecutive days, a juvenile flew from one of these nests when a researcher arrived to weigh the nestlings, to return after he had left. It is not always clear whether the young Wood Pigeon fed by parents of nestlings was their own progeny from an earlier nest or an unrelated bird. A fledgling Domestic Pigeon, about twenty-five days old, who had begun to pick up corn but was still being fed by hand, regurgitated food to a younger companion, in the manner of an adult feeding its young. Probably the young helper had no crop milk to give, but the recipient was old enough to digest more solid food.

At three among nearly five hundred nests of Mourning Doves, David Blockstein found two males, each of whom sometimes incubated the eggs and brooded and fed the nestlings, without taking equal shares in attendance. At none of the three nests did the female show a preference for one of her assistants. The two males fought together, striking each other with their wings. Apparently, at each of the three nests, an unmated male tried to participate in the activities of an established pair, but his help was not welcomed by the mated male.

Among interspecific helpers, a Mourning Dove in southern Arizona brooded and fed nestling White-winged Doves whose own parents had neglected them much of the day. For reasons unknown, the father of these nestlings never appeared. When at last the mother White-wing returned, late in the afternoon, she fought and drove away the helpful Mourning Dove. Nevertheless, the latter continued to attend the young White-wings, whose mother was not again seen, until they fledged. Probably, the helper had laid eggs that had failed to hatch. In aviaries, Mourning Doves and other pigeons readily adopt and aid in the care of young pigeons of other species.

Pigeons, like cuckoos, are unskilled builders that occasionally deposit their eggs in nests of other birds, probably because they could not finish their own in time to receive them. A century ago, a Yellow-billed Cuckoo laid two eggs in an unfinished nest of an American Robin, in which the robin also laid an egg. Then a Mourning Dove added two eggs to the mixed set and incubated beside the cuckoo. The discoverer of this extraordinary situation thought it more important to collect the eggs than to learn the natural outcome.

In two consecutive years, an American Robin and a Mourning Dove laid eggs in a robin's nest in precisely the same site. Here they incubated alternately, thereby becoming mutual helpers. Each would watch from a nearby branch until the other left the nest, then go to take its turn on it. Neither was antagonistic to the other. The mates of these two females were not found. In the first year, boys destroyed the eggs before they hatched. In the second year, two eggs of each collaborator hatched, and the young were fed and brooded for eight days. When Edward Raney, who found these nests, approached the parents closely, each behaved in a characteristic way, the robin scolding loudly from a nearby limb, the dove fluttering over an adjoining field in a broken-wing display. By the ninth day, all four nestlings had died in the nest from an unknown cause. In view of the very different ways that pigeons and thrushes feed their young, it is remarkable that the mixed brood survived so long. Did each parent nourish only her own offspring?

8 Rate of Reproduction

The farther from the inner tropics (within about twelve degrees of the equator) that one goes, the more eggs one is likely to find in a bird's nest. This "latitude effect" is exhibited by many families and even by species of wide latitudinal distribution. No pigeon is known regularly to lay more than two eggs, and a large minority lays only one. It is significant that most single-egg species occur in the tropics and subtropics, although a few—for example, the Band-tailed Pigeon in the western United States—range from the tropics, where they are most widely distributed, to fairly high northern latitudes. The most widely distributed pigeons in the temperate zone of Eurasia are the Rock Dove, the Eastern Rock Dove, the Stock Dove, the Wood Pigeon, the Turtle-Dove, the Eastern Turtle-Dove, and the Collared Dove. The only native pigeon widely distributed in the temperate zone of North America is the Mourning Dove. All these northern pigeons lay two eggs. The extinct Passenger Pigeon, which laid only one, was, by any criterion, a most exceptional member of its family. In Australia, many pigeons that range well south of the Tropic of Capricorn have two-egg clutches; but at higher austral latitudes, the New Zealand Pigeon and the Chilean Pigeon incubate single eggs.

In the tropics, pigeons live among many other birds that lay no more eggs than they do. Clutches of two are very frequent, and birds not a few, among cotingas, sunbirds, birds of paradise, nightjars, and potoos, incubate only one at a time. In the North Temperate Zone, most of the pigeons' neighbors produce larger sets of eggs, often three, four, or more. Why, then, do these northern pigeons rear such small broods? Can it be that the female is unable to collect, in the available time, enough food to form more than two eggs? Is the limiting factor the parents' ability to warm more eggs? Is a pair unable to nourish adequately more than two young? Or is the nest inadequate to hold more than two?

Pigeons' eggs are exceptionally small in relation to the size of the bird who lays them. Those of the Wood Pigeon weigh about 20 grams, the set of two about 40 grams, which is only about 8 percent of the weight of an adult female. Other birds, with larger sets of relatively heavier eggs, manage to mobilize enough material to produce clutches about equal to their own weight. Courtship feeding by male pigeons helps the females to form their eggs. For the shells, at least some pigeons eat many snails, whose shells supply calcium, or they visit the salt licks of cattle to supplement their intake of minerals. Moreover, after feeding one brood, female pigeons can in a few days set aside enough nutrients to produce another set of eggs. It is not obvious that the female's ability to produce eggs limits the number she lays.

We earlier learned that a pigeon's nest sometimes contains three or even four eggs, the product of more than one female. Parents are well able to incubate and hatch at least three eggs. Long ago, C. V. Duff watched Mourning Doves attend triplets in California. A pair of Galápagos Doves also reared a brood of three. Among 3,043 eggs in a huge colony of Eared Doves in Argentina were 520 in sets of 3; 34 percent of these eggs hatched, about the same proportion as in nests with only one egg. Predation, rather than failure to hatch, was the principal cause of loss. By shifting eggs of equal age among nests of Mourning Doves to make sets of one and three, David Westmoreland and Louis Best demonstrated that parents had no difficulty incubating the augmented sets. All fertile eggs not eaten by predators hatched successfully, regardless of the number in a nest.

Not only can Mourning Doves and Wood Pigeons hatch three nestlings, they can brood and feed them. Birds of diverse families increase the amount of food they bring to a nest as the number of nestlings increases, but not proportionately, so that a solitary nestling receives more than a member of a brood of two, and each of the latter does better than a member of a brood of three, even when this brood is not abnormally large. The same is true of pigeons, with the result that three young Mourning Doves grew more slowly than two while they were fed chiefly with milk, and their crops were usually less full throughout the nestling period. Enlarged broods of Mourning Doves took, on average, 1.3 days longer to fledge and weighed about a quarter less than control broods. Nevertheless, nests with three young Mourning Doves were slightly more successful than those with only one and, inexplicably, much more successful than those with the normal brood of two. In an experiment with Wood Pigeons conducted by R. K. Murton, 84 percent of ninety-nine chicks raised as triplets survived to fledge, whereas 98 percent of those in broods of one and two lived to leave their nests. However, the triplets did not survive so well after fledging, whether because of inadequate parental care after they flew or for other causes is not clear. Passerine birds who are underweight when they fledge can often compensate for their deficiency after they leave the nest and follow parents who feed them.

Perhaps pigeons limit the size of their broods to one or two because their nests cannot safely accommodate more young. Westmoreland and Best

Croaking Ground-Dove *Columbina cruziana,* male
Also called Gold-billed Ground-Dove.
Female duller. Arid and semiarid Pacific
coast of South America
from Ecuador to northern Chile.

found that nestling Mourning Doves in enlarged broods often fell or were pushed from their nests. To avoid this source of loss while they tested the parents' ability to nourish triplets, they set the nests in camouflaged wire-mesh cones, securely attached to branches at or near the original sites. By this means, they increased the success of their enlarged broods. This raises a question that also arises when we consider the nests of hummingbirds, manakins, certain cotingas, and other birds whose nests are barely ample enough to accommodate a brood of two, or sometimes only one. Do these birds rear small families because they cannot build nests capacious enough for more young, or do they build small nests because their broods are small? Does the size of the brood control the size of the nest, or vice versa?

Although some ornithologists have contended that the broods are small because the nests are small, I believe that the opposite is true. If it were advantageous for these birds to rear larger families, they would build bigger or more substantial nests. Breeding is commonly a strain on parent birds, reducing their chances of survival. By raising smaller families, less costly of energy and effort, they may increase their life expectancy and leave more progeny in the long run. Moreover, excessive reproduction may penalize rather than benefit a species. By overburdening its resources and causing it to suffer high mortality during the leaner months of a year, excessive reproduction reduces its population below the point that might be maintained with more restrained fecundity.

Pigeons compensate for their small broods in several ways. Many species have prolonged breeding seasons. In arid southwestern Ecuador, S. Marchant found nests of the Croaking Ground-Dove in all months except September and October. In northern South America and on the neighboring island of Trinidad, Ruddy Ground-Doves nest throughout the year. Yearlong breeding is also reported of the White-tipped Dove in Trinidad. In the Costa Rican valley where I write, at the same latitude but at a higher altitude, about twenty-five hundred feet (760 m), this dove has two shorter breeding seasons, the principal one from December (rarely) or January into April, with reduced nesting from July to October. Strangely, no nests of White-tipped Doves or of their neighbors in semiopen country, Gray-chested Doves and Blue Ground-Doves, have been found in May and June, when breeding of the avifauna as a whole is at its height. In the rain forest, however, Ruddy Quail-Doves nest continuously from March to August, as they do during the same interval in other parts of their far-flung range, in northern South America and on the island of Jamaica.

Beyond the tropics, Eared Doves in the province of Córdoba, Argentina, nest throughout the year, but most abundantly from January to April—the austral summer and autumn—with a marked decline in the winter months of June and July. In California, Mourning Doves have been found nesting in every month except October and November, but mainly from March to September. In Texas, nesting has been recorded from late February through September; farther north and east, from Iowa to New England, the breeding season extends from April to August or September, with few eggs laid earlier or later. In the southern United States, Common Ground-Doves and Inca Doves lay their eggs from late February through October. In England, Wood Pigeons have been found nesting throughout the year, but mainly in the warmer months, from May or June until October. In the grain-growing districts of Cambridgeshire, nests are most abundant late in the summer, when the crops on which Wood Pigeons largely depend begin to ripen. The pigeons who try to rear their families earlier have poor success because they often leave their nests exposed while both parents seek scarce food, with the result that many of the nests are pillaged by jays, magpies, and other corvids.

By refurbishing a nest that remains intact and using it over and over, pigeons can raise their broods in rapid succession. A successful nesting by a pair of Mourning Doves occupies about thirty days; thus, in six months of favorable weather, as in California and Texas, they have time to rear six broods. Because their nests are often lost, however, few produce more than three. One exceptionally successful pair in California did rear six broods, each of two fledglings, all twelve in the same nest. In Iowa, with a slightly shorter breeding season, a few nests were used five times, but in none were more than four broods raised. Wood Pigeons in England have time to rear three broods but achieve this result only rarely, in very favorable years and in well-protected situations. Only a small proportion of the population succeeds in rearing two broods. In Germany, Wood Pigeons, Rock Doves, Stock Doves, and Collared Doves raise two or three broods; the Turtle-Dove, two. In the Kenya highlands of East Africa, a pair of Red-eyed Doves nested five times in twelve months, and each of two other pairs built four nests. Among other tropical pigeons, Ruddy Ground-Doves in Trinidad occupied a nest for five consecutive broods; how many were successful is not stated. I have known White-tipped Doves to start three broods, two of which successfully fledged. Ruddy Quail-Doves sometimes lay twice in the same nest.

The short interval between nestings, often in the same nest, no less than the rapid development of the young, helps pigeons to nest repeatedly in the same season. From one to six days after a brood leaves the nest, Mourning Doves commonly start to lay again, unless the favorable season is ending. Following a nesting failure, the interval is more variable. A female who has lost her eggs or nestlings may start a new clutch from two to twenty-five days later, but most often she does so in three to six days. Occasionally, a Mourning Dove shortens the interval between broods by laying an egg in a nest that still contains young of the preceding brood, about ready to depart. Wood Pigeons in England may similarly resume laying while still attending nestlings.

Among captive pigeons supplied with abundant food, nest sites, and nesting materials, overlapping broods are much more frequent than among free pigeons, or at least more often noticed. The Feral Pigeon's long nestling period of twenty-five to thirty-four days is especially conducive to overlapping broods, at least when these pigeons are kept in aviaries. In Texas, Nancy Burley learned that pairs of fully experienced Feral Pigeons were simultaneously engaged in rearing young and incubating eggs 70 percent of the time. Pairs consisting of one partner who had already nested and an inexperienced mate overlapped their broods less often; pairs nesting for the first time overlapped only rarely.

The most exacting interval in a pigeon's nesting cycle is the first days after hatching, when the nestlings are nourished on crop milk. After the parents begin to feed their nestlings with grains, their task becomes lighter and they can begin to prepare for another brood. When the first brood of Feral Pigeons is about two weeks old, their mother lays a new set of eggs. The male is now the chief attendant of the half-grown nestlings, but he nevertheless incubates for three to six hours a day. The eggs hatch when the first brood is ready to leave the nest. The parents now have two sets of young to feed; but after a week or so, the fledglings are becoming independent. Burley's study suggests that the supply of milk during the nestlings' first week is the major constraint on the size of pigeons' broods.

Although overlapping broods have been reported for a number of birds of other families, they appear to be far from common. When they occur, the male parent usually feeds the young while the female alone incubates the new set of eggs; but with the pigeons' schedule of nest attendance, this would not ordinarily occur. It would be interesting to know how free pigeons arrange matters.

The reproductive potential of some pigeons is increased by precocious sexual maturity. Common Ground-Doves in southern Texas were ready to breed when only seventy-nine days old. Mourning Doves in Arizona can nest at the age of three months, and captive Eared Doves at four months. In these species, the young can contribute substantially to reproduction in the same year that they hatch. Few birds of other families are ready to breed before they reach or approach their first birthday, and many wait several years before they nest.

Small birds with open nests lose a high proportion of their eggs and nestlings. For several reasons, losses are nearly always higher during the egg stage than during the nestling stage. Nests that survive until the eggs hatch have already escaped many perils. They tend to be better made, better concealed, or less accessible to predators than those that are lost early. Also, parents are more strongly attached to their young than to their eggs and are readier to defend them. Table 4 summarizes the nesting success of nine species of pigeons in diverse habitats. The outcomes of nests, even of a single species, vary so greatly with locality, season, and presence or absence of particular predators that the results of the most careful studies are at best rough approximations of how a species fares over a wide area. An inspection of the table suggests that approximately half of all pigeons' nests in which eggs are laid survive to produce at least one fledgling, which is about average for small birds with open nests in the North Temperate Zone. Because a nest is counted as successful even if the number of fledglings it produces is less than the number of eggs it contained, the percentage of eggs that yield fledglings is usually lower than that of successful nests.

I find it interesting that Feral Pigeons nesting on the docks, warehouses, flour mills, and other buildings in an industrial area of the inland port of

Common Ground-Dove *Columbina passerina,* male
Southern United States to Ecuador and eastern
Brazil; Bermuda, Bahamas, and Greater and Lesser Antilles.

Table 4. Nesting success of pigeons

Species	Nests with eggs	Successful nests (number and percent)	Eggs laid	Successful eggs (number and percent)	Locality
Feral Pigeon			1224	638 (52)	Manchester, England
Wood Pigeon			1704	528 (31)	England
White-crowned Pigeon	170	111 (65)			Puerto Rico
Mourning Dove	940	600 (64)			Illinois, United States
Mourning Dove	3439	1687 (48)	6427	3085 (48)	Iowa, United States
Mourning Dove	200	142 (71)	398	274 (69)	California, United States
Eared Dove	1126	524 (47)	2300	752 (33)	Argentina
Galápagos Dove	56		110	79 (72)	Galápagos Islands
White-winged Dove			17,524	5547 (32)	Texas, United States
Croaking Ground-Dove	283	160 (57)	477	298 (62)	Ecuador
Ruddy Quail-Dove	17	5 (29)	33	10 (30)	Costa Rica

Note: The table is not uniform because authors have presented their data in diverse ways.

Manchester, England, reproduced about as successfully as Mourning Doves in small towns, farmyards, and patches of woods in agricultural Iowa. Their nests were more successful than those of Wood Pigeons scattered over the British countryside. Mourning Doves nesting in the willows surrounding a small pond and in trees around the headquarters of a wildlife refuge in northern California enjoyed higher-than-average success. Likewise, Croaking Ground-Doves in arid southwestern Ecuador produced flying young in a high proportion of their nests. In tropical rain forests, where nests of most kinds of birds are widely scattered and well hidden, it is difficult to find enough for statistical comparisons. My small sample of rain forest nests of Ruddy Quail-Doves was much less successful than the nests of other pigeons in more open places, but 30 percent is about average for open nests in rain forest.

We wish to know how well nests succeed when undisturbed by human visits, but obviously we cannot learn their outcomes without visiting them, and this introduces a factor difficult to assess. Numerous studies to determine the effect of human disturbance on the productivity of birds' nests, conducted with a diversity of species, have yielded contradictory conclusions. Some have indicated that visits of inspection, made with a minimum of disturbance, do not adversely affect nesting success. Others have shown that the investigators' approaches to nests increase the probability of predation, possibly by leaving scent trails that keen-nosed quadrupeds can follow, or, in the case of uncamouflaged eggs such as those of pigeons, leaving them exposed to view until a parent returns to cover them. When Murton and his co-workers examined Wood Pigeons' nests at weekly intervals, about 68 percent of the eggs were lost to predators; but when inspections were made fortnightly, losses were reduced to 55 percent. Hatching success was much greater in high than in low nests, because parents did not fly off when watched from below and their eggs were less often left unguarded.

In Iowa, Westmoreland and Best compared the outcome of Mourning Doves' nests that were checked from a distance, without putting the parents off their eggs or young, with those from which the adults were driven off, at three-day intervals, for a closer inspection of the nests' contents. The undisturbed nests were twice as successful as the disturbed nests. However, in Maryland, James Nichols and his associates found no significant difference in the outcomes of Mourning Doves' nests that were visited at daily and at weekly intervals; in both cases, the parents were put off so the nests' contents could be examined. On the whole, it appears that consequences difficult to avoid in investigations of breeding success tend to increase losses, but if visits of inspection are limited to the minimum necessary to learn when eggs are laid, if and when they hatch, and if and when the young fledge, their effect may be minimal.

In the North American midwest, gales that blow eggs from nests or nests from trees are a major cause of losses by Mourning Doves. Hailstones break eggs and kill parents and young. Mourning Doves diminish some of these losses by using the more substantial nests of robins and grackles as bases for their own slight structures. Nests of these doves are not all equally fragile; Richard Coon and his collaborators learned that those built well enough to remain intact over winter were more successful than those that vanished or, if still present, were too dilapidated to hold eggs. Likewise, nests in forks of trees tend to do better than nests supported on branches.

In the North Temperate Zone, birds of the crow family, including jays, magpies, and grackles, eat many eggs and nestlings; squirrels, chipmunks, raccoons, weasels, and other small arboreal mammals add to the destruction. In Puerto Rico, Pearly-eyed Thrashers are major enemies of nesting pigeons and other birds. Snakes appear to be the chief destroyers of birds' nests in many regions, above all in tropical woodlands, where these reptiles abound. On the island of Guam, introduced brown tree snakes ravage so many nests that they threaten the extinction of the Javanese Collared Dove and other long-established and native birds.

Although, as told in chapter 7, pigeons valiantly defend their nests against snakes and other intruders, they risk being swallowed if the serpents are big; and on dark nights, they are at a great disadvantage when confronted even by small ones. When pigeons are suddenly alarmed, as by stampeding cattle, running horses, or noisy children, they frequently knock eggs or young from their nests by panicky departures. With so many sources of loss, many pigeons need their high fecundity merely to maintain their populations. Others, especially on small islands where they faced few threats before man arrived with his satellite animals, have been unable to reproduce rapidly enough to avoid extinction. Still others, usually in continental environments greatly altered by man, have become excessively prolific, with unfortunate consequences for themselves and for humans—a subject we will explore in the following chapter.

Pigeons are rarely troubled by parasitic cowbirds, whose young are reared to the detriment of the foster parents' own nestlings. Pigeons' widespread practice of keeping their nests constantly covered after the first egg is laid makes it difficult for a cowbird to foist her eggs on them. Among the few reports of cowbirds' eggs in pigeons' nests are those of the Brown-headed Cowbird in nests of the Mourning Dove and the Common Ground-Dove; of the Shiny Cowbird in nests of the Eared Dove, the Picui Dove, and the Black-winged Ground-Dove; and of the Bronzed Cowbird in nests of the White-winged Dove, the Mourning Dove, and the Common Ground-Dove. Pigeons are hardly likely to rear young cowbirds, because the foods of pigeons and passerine birds such as cowbirds and the methods of delivering them are very different. A nestling pigeon inserts its mouth into that of a parent; a nestling cowbird lifts its gaping mouth for a foster parent to place food into it. Despite this difficulty, a Mourning Dove is reported to have reared a Brown-headed Cowbird on at least one occasion.

9 Pigeons and Man

In the Metropolitan Museum of Art in New York City is a marble gravestone, from the Aegean island of Paros, carved in the middle of the fifth century B.C. with the meticulous attention to detail characteristic of the best Greek sculptures of the Classical period. The high relief depicts a lovely young girl holding two pigeons. One perches on her left arm, head turned toward her. The other, pressed against her breast by her right hand, appears to be taking food from her lips, or perhaps receiving a kiss. Gazing, not without emotion, at these lifelike figures that have so well escaped the ravages of time, I tried to imagine how the girl lived and felt. Surely, like many another girl or boy of more recent ages, she loved her pigeons dearly. Did she give them names and murmur softly to them in the language of Sophocles and Plato? Did they come to her call, for food or caresses? Did they survive her, or did she grieve for them before her own early death, as her parents must have grieved for her? This ancient memorial, commissioned by bereaved parents, reminds us that our intimate relationship with pigeons is very old, and frequently charged with deep feeling.

People's feelings about the animals that share Earth with them range from fervid protectiveness to delight in their destruction and greed for their flesh. In no context is this more clearly evident than in the attitudes of the inhabitants of temperate North America toward the only pigeon widespread in their countries, the Mourning Dove. In about a third of the contiguous United States, chiefly in the Midwest and Northeast, the Mourning Dove has been fully protected as a songbird; in the remainder, it has been classed as a game bird with an open hunting season. In 1984, Indiana, against the heated protests of many of its citizens, withdrew the protection that for 111 years it had given these doves. The opportunity to shoot them was gleefully celebrated by Ted Williams in an article in *Audubon Magazine* called "The Quick Metamorphosis of Indiana's Doves." Since Mourning Doves are so numerous and many people enjoy shooting them, he argued, why should they not be shot? Many readers were so offended by this tactless article, by its levity of tone, and by its inadequate attention to the doves' breeding system and ecology that they wrote angry protests to the magazine, canceling their subscriptions, and threatening to resign from

Mourning Dove *Zenaida macroura*
Sexes alike. Southern Canada to central Panama; Bahamas and Greater Antilles.

the National Audubon Society, which publishes it. They could not understand (as I cannot) how anybody could take pleasure blowing the life out of a beautiful, inoffensive dove—or any other harmless animal. Sensitive, thoughtful people find higher values in watching living pigeons, with their evident affection for their mates and young. Among them are Australians, who have given full legal protection to all twenty-two of their indigenous species.

Probably because the Rock Dove nests in sheltered crannies and on buildings, instead of in the open like most pigeons, it became associated with man soon after he began to build permanent dwellings and religious edifices. Its domestication has been traced to the Bronze Age in the Middle East, over six thousand years ago. Here it was regarded as sacred to Astarte, or Ishtar, goddess of love and fertility. Similarly, the Greeks associated the pigeon with Aphrodite; the Romans, with Venus. In India, Kamadeva, the Hindu equivalent of Eros or Cupid, rode on a dove, bearing in one hand an arrow made of flowers and in the other a bow whose string was formed by bees. Probably the affectionate "billing and cooing" of doves, their fertility, or both together, caused them to be chosen as symbols of love and procreation.

Pigeons early acquired a reputation for faithfulness and dependability and for being a vehicle of spiritual values. The story of how the dove that Noah released from his ark brought back an olive leaf as evidence that the Deluge was subsiding is too well known to need repetition. This account, like much else in the Bible, is based on the lore of older cultures near Palestine. In one version of the story of the Deluge, the dove returned to the ark with muddy red feet, proof that some land had emerged from the receding waters. Noah prayed that pigeons' feet might forever retain this color, as indeed they have in many species throughout the world.

In the Gospels of Matthew and Mark, we are told that, as Jesus rose from the water in which he was baptized, he saw the Spirit of God descending on him in the form of a dove. Probably as a consequence of these passages, the dove became an emblem of the Holy Ghost in early Christian art. Issuing from the lips of dying saints and martyrs, it represented a blessed soul's release from the body. Ironically, in Christian countries where today the dove is widely regarded as a symbol of purity, love, and peace, doves and pigeons are slaughtered on a vast scale, as at the battues in which an army of gunners blasts away at the birds as they return at nightfall to their communal roosts. In Moslem lands where the tradition of their sacredness lives on, pigeons are protected and fed, especially around the mosques.

Pigeon-keeping spread through the early Mediterranean cultures, as the sculpture of the Parian girl with her doves attests. Aristotle could give a good account of the reproductive behavior of pigeons, probably based largely on observation of domestic breeds. The practical Romans took great interest in raising pigeons and knew their value as winged messengers, as other nations had apparently done before them. While conquering Gaul, Julius Caesar employed them to carry dispatches; and in the civil wars that followed his assassination, in 44 B.C., generals communicated by means of pigeons. Through released homing pigeons, news of victories in athletic contests and chariot races was speedily conveyed to families eager to learn the outcomes. Romans also raised pigeons as a hobby, paying high prices for them and recording their pedigrees. Probably they already had some of the fancy breeds. In the Middle Ages and early modern times, manorial estates throughout Europe maintained dovecotes or pigeon lofts as sources of food.

Supplementing the ability of pigeons to spread unaided throughout the tropical and temperate zones, adapting themselves to the most diverse climates, man has carried certain of them far and wide. Above all, Feral Pigeons, more like the ancestral Rock Dove than are many of the domestic breeds, have by his agency become residents of many lands and are familiar sights in innumerable cities and towns from the tropics to the Arctic Circle. Although probably not so well fed as in the days when horses, widely used for traction in cities as well as on farms, spilled some of their rations of grain, urban pigeons continue to thrive. Their presence in large numbers is not without undesirable consequences, to which we will presently return, but many people whose contacts with nonhuman animals are few find pleasure in watching and feeding them. If they could remain more consistently in parks and open places, they would be more consistently welcomed; but they need buildings for nesting, often to the detriment of these structures or of the people who occupy them.

The Ringed Turtle-Dove, in England called the Barbary Dove, is a long-domesticated form of the African Collared Dove which has escaped from captivity and become established at widely separated points in the extreme south of the United States, including localities in Florida, Alabama, Texas, Arizona, and California, and in Puerto Rico and the Bahama Islands. Turtle-Doves live chiefly in cities and suburbs where people supply food, and the feral population is reinforced by occasional escapes from captivity.

The Spotted Dove, native of southeastern Asia and Indonesia, has been introduced and become established in coastal southern California, Hawaii, Mauritius, Australia, New Zealand, and several islands of Polynesia. In most parts of its range it is found in or near cultivated areas. Its habit of feeding along roads or paths in towns and villages costs the lives of many as automobiles become common where they were previously rare or absent. Although Old World pigeons, notably the Rock Dove, have with man's aid successfully invaded the New World, the reverse appears not to have occurred. As far as I can learn, no New World pigeon has been naturalized in the Old, as have the Canada Goose and a number of mammals, including the American gray squirrel and the muskrat.

Although man has helped a number of species to expand their ranges or increase their populations, directly by importation and indirectly by converting woodlands into grain fields and pastures where certain pigeons forage, he has reduced the ranges and populations of others, too often to zero. The most spectacular and publicized case of the extinction of a pigeon is that of the Passenger Pigeon, which in eastern North American inhabited Earth's most extensive broad-leaved temperate-zone forest in such astronomical numbers that it was perhaps the world's most abundant bird. The destruction of these forests on a vast scale during the eighteenth and nineteenth centuries inevitably reduced the population of this pigeon, but it might well have survived in smaller areas of woodland if it had not been slaughtered insanely by unregulated market hunting.

The other six pigeons that have become extinct since 1600 inhabited small islands in the Pacific and Indian oceans. Numerous factors have contributed to the extinction, throughout the oceans, of an appalling number of insular species of pigeons and other birds: the original smallness of their populations, compared with those of continental species; the introduction by man, accidentally or deliberately, of rats, cats, pigs, and other predatory mammals, on islands long devoid of such threats to birds and their nests; the advent of mosquitoes that spread avian malaria and other diseases among birds without resistance to them, as on the Hawaiian Islands; and, finally, direct persecution by man, often on islands that had never supported humans, such as New Zealand, where the moas were hunted to extinction by the Maoris. The flightlessness of many island birds, such as the Dodo, increased their vulnerability to invaders, human or otherwise, of their isolated homes.

In addition to the seven species of pigeons that have become extinct in the last four centuries, sixteen more are listed as endangered. Among them is the lovely Pink Pigeon of Mauritius, which, by the reintroduction of birds bred in captivity, the study of its habits, and conservation measures may still be saved from sharing the unhappy fate of fourteen other birds that once lived with it on the Mascarene Islands in the Indian Ocean, including the Mauritius Blue Pigeon, the Dodo, and the solitaires. Many of Australia's pigeons have become much less abundant since Europeans arrived two centuries ago.

A few species of pigeons have been stigmatized as pest birds. The most widespread of the pigeons that in some places have become annoyingly abundant is the Feral Pigeon, which today thrives in many places that it would not have reached without human aid. In cities, it fouls the buildings on which it nests and roosts, sometimes causing the deterioration of architectural adornments and costly cleaning operations. It dirties sidewalks, with unpleasant consequences for pedestrians, and enters warehouses and food-processing plants for grain. Pigeons, along with European Starlings and other birds that in large numbers roost or nest in cities, are often regarded as hazardous to human health because they are vectors of psittacosis, or chlamydiosis, a disease that in man takes the form of a viral pneumonia. Histoplasmosis, an infection that produces enlargement of the human liver and spleen, fever, and anemia, results from inhaling spores of

Feral Pigeon *Columba livia*
Derived from domesticated Rock Doves. Cosmopolitan.
House Sparrow *Passer domesticus,* male
Originally Eurasia and northern Africa,
now cosmopolitan.

the fungus *Histoplasma capsulatum,* which grows on bird droppings. However, the probability of contracting these diseases or infections is slight unless one is in prolonged and intimate contact with birds, as in aviaries and commercial breeding establishments. In any case, chlamydiosis, which is potentially lethal, is readily cured by antibiotics; and 95 percent of people who are exposed to histoplasmosis develop immunity without suffering unpleasant symptoms. Moreover, the fungus that causes histoplasmosis appears to be present in all organic soils; all garden soils in the Ohio and Mississippi valleys are infected with it. If every last bird were eliminated, the fungus would still be with us.

One method of reducing the numbers of urban pigeons is to construct strictly functional, straight-sided buildings, without the traditional architectural embellishments rich in ledges and recesses where the birds roost or nest. Since shooting is generally prohibited in cities, other means of decimating pigeon populations have been widely employed. Pigeons have been trapped, and they have been immobilized by ingesting grain treated with a stupefying chemical, such as alpha-chloralose. The ideal method of holding populations, whether of humans or other animals, within sustainable or endurable limits is to depress reproduction by harmless treatments. For any creature, it is better never to have been born, or hatched, than to live miserably and die prematurely of slow starvation or diseases induced by stress.

Success in controlling the screwworm infestation of cattle by the release of large numbers of male flies artificially reared and sterilized by radiation prompted researchers to seek a means to repress the reproduction of problem birds. Birds were sterilized by means of food treated with an appropriate chemical. Because it is hardly possible to regulate the doses ingested by free birds, some took too much and suffered debilitating side-effects, including death; others took too little. The best results were obtained by holding the birds captive long enough to hand-feed them sublethal doses in capsules over successive days, until a sufficient amount had been administered—a procedure both time-consuming and costly.

Moreover, it was found that in certain populations of urban pigeons, as in Manchester, England, about two-thirds of the adults do not attempt to breed in any case; sterilizing them would thus be so much misspent effort. After a prolonged study of pigeon control at the Manchester docks, where enormous quantities of grains are imported and processed, and a careful balancing of the costs and benefits of removing pigeons, R. K. Murton and his co-workers recommended a different approach to the problem. Instead of methods of control that are both expensive and ineffective, they advised taking measures to prevent or clean up the spillage of the grains that support the present populations of Feral Pigeons. In addition to bringing other benefits, port hygiene would ameliorate a pigeon problem not readily solved by other means.

In England, another problem bird is the large Wood Pigeon, which in grain fields and market gardens may cause large losses. As Murton pointed out, farmers (and others) tend to exaggerate their losses or to cry prematurely that their crops have been ruined. Thus, a field of clover that in winter has been heavily grazed by pigeons may, by explosive growth in March, produce in May or June as much as a field that has remained intact. Similarly, a plot of cabbages that appeared to be a total loss after severe winter grazing by Wood Pigeons did yield a good crop, thanks to the plants' great recuperative power, but too late in the season to bring the highest prices. An experiment showed that plots of peas thinned by pigeons during the seedling stage yielded, on average, as much as plots screened to keep the pigeons out. By growing larger and bearing more pods, the thinned peas compensated for their fewness. In other cases, pigeons caused great losses to individual farmers, but to the nation as a whole the loss was less than the cost of controlling the birds. The effect of shooting was to kill birds that in any case would die of other causes; it did not regulate the size of the pigeon population. "If any lesson can be learned from this account," wrote Murton, "it is perhaps that sport and serious pest control are not necessarily compatible." He recommended, as deterrents to the depredations of Wood Pigeons, anchoring numerous hydrogen-filled balloons over grain fields subject to their attacks and shooting the birds only at times and in places where they were causing losses, when it would also scare them away.

The Eared Dove lived for generations as an innocuous member of the Argentine avifauna. In the middle of the present century, however, it rather suddenly came into prominence as a menace to agriculture, with a great increase in population, especially in the province of Córdoba. Roosting and nesting by the millions in areas of scarcely penetrable thorny scrub, the doves spread widely over surrounding fields sown with millet, wheat, sunflower, and sorghum (chiefly the last-mentioned), often devouring a substantial proportion of the yield. After considering the ineffectiveness of shooting as a method of controlling the numbers of such prolific birds, and the dangers of widespread poisoning to man, other animals, and the environment, Enrique Bucher proposed another solution to the problem, essentially protecting the crop rather than trying to eliminate the doves. The grain could be harvested early, while its water content was high, then dried artificially. In the field, drying might be accelerated by harvest-aid chemicals. Resistant varieties of sorghum would greatly reduce losses to birds. Repellents might keep them away. The doves should be destroyed only when and where they were damaging crops. Moreover, less wasteful harvesting equipment might reduce losses caused by spilled grain, which often exceed those caused by birds and other animals; this spilled grain also helps support high concentrations of the doves.

In Texas and Arizona, White-winged Doves flock to fields of ripening sorghum and other grains, in some years and in some places causing substantial losses to farmers, although their depredations appear not to be as serious as those of Eared Doves in Argentina. Scaring the birds away with various noise-making devices and patrolling the fields as the grains ripen have been preferred to killing the birds. By planting their sorghum early, growers are often able to harvest it before August, when the doves descend on the fields in the greatest numbers. Also, when such native food plants as doveweed and leatherweed seed heavily, in pastures and uncultivated fields, they attract the White-wings away from sorghum.

Only a small proportion of Earth's nearly three hundred species of pigeons are troublesome to man, and these only in special situations, often of man's making. Some are welcome neighbors in our dooryards and shade trees, where by their gentle ways and soft voices they help create an aura of peace and tranquillity in a world where these blessings have become increasingly rare. Others, less approachable, hold our delighted gaze as they perch majestically at the very top of a towering tree or as they rise before us along some solitary woodland trail. Some are among the most beautiful of birds; few lack adornments, if only on their iridescent napes or their wings. Their family life is exemplary; both sexes participate in all domestic tasks with an equality of effort that we might envy. People have long loved pigeons, cherished them, found pleasure and comfort in their companionship, and still find them useful for carrying messages and in other ways. At the other extreme, they have mercilessly persecuted them, slaughtering vast numbers, exterminating whole species, subjecting them to cruel experiments. They deserve better treatment from rational beings who have long esteemed them as symbols of love, peace, and purity.

10 Homing Pigeons

Man has domesticated a diversity of birds and mammals as sources of food, and of quadrupeds to transport him and his burdens, but the only animal he has widely reared as a messenger is the homing pigeon, a descendant of the Eurasian Rock Dove. When we recall that the use of birds to carry messages implies the ability to orient to definite points from considerable distances, the choice of the Rock Dove for this service is surprising. The wild Rock Dove, whose travels are usually no farther than between roosting or nesting cliffs and foraging areas, does not exhibit the navigational skill or ability to fly long distances of hundreds of other avian species, including a few pigeons, which twice yearly travel over vast expanses of land and water. But animals of many kinds have an ability to find their way over great distances, including those never traversed by them. This ability reveals a closer rapport with Earth and its many emanations, a greater sensitivity to tenuous stimuli, than civilized man has retained from remote ancestors who probably were not so obtuse.

From a measure of homing ability undoubtedly present in the earliest domesticated stock, breeders have by careful selection and training produced pigeons that can return home from afar. The establishment of these pigeons in lofts and dovecotes that substitute for their ancestral caves and crevices in cliffs makes them easier to rear, and to recover when they return from a journey, than would be possible with pigeons that live in trees.

The preceding chapter told how pigeons were employed by ancient Greeks and Romans to announce victories in athletic contests and chariot races. More important was their use as messengers by warring armies, and probably also in routine civil administration, for until the relatively recent inventions of the telegraph, telephone, and radio, they were the most rapid means of long-distance communication available. In Egypt and through the Middle East, their homing ability was recognized and widely used. It is related that, in the sixteenth century, Akbar, the "Great Mogul" of northern India, maintained large numbers of pigeons, probably for keeping informed of events throughout his empire and communicating with neighboring potentates. Pigeons proved to be especially helpful to cities closely invested for long periods by armies that severed all other means of communication with the outside world, as in the siege of Acre by the Crusaders, the six-month siege of Leyden by the Spaniards during the Netherlands' War of Independence, and the siege of Ladysmith in the Boer War in South Africa.

The most famous of all pigeon posts was established during the Prussian siege of Paris, from November 1870 to January 1871. Parisian pigeons were placed in balloons that bore them safely over the besiegers' lines. They were then taken to London, Tours, and other cities, where they were released; about 17 percent of them returned to their home lofts with messages. At first, the messages were written on paper, rolled up tightly, covered with wax, and fastened to a tail feather. Many of these letters were lost. Later, as many as twenty-five hundred were microphotographed on a film so thin that a number of pieces could be carried by a single bird; the films were inserted in a small goose quill tied with waxed silken thread to the strongest feather of a pigeon's tail. One pigeon carried about forty thousand messages on eighteen of these tiny films. By the time the siege was lifted, 150,000 official and a million private communications had been delivered to Paris by the winged postal service. Its success prompted several European countries to provide carrier pigeons for their armies.

Even in the two world wars, when the telegraph and other modern methods of communication were available, homing pigeons proved of value, for they could carry messages when wires were cut, radio transmitters were unavailable or out of order, and men or dogs were unable to pass alive through the enemy's lines. Recognizing the value of pigeons, in World War I the Germans promptly ordered the destruction of all pigeons in the territories they occupied in Belgium and northern France. Moreover, they are said to have confiscated about a million Belgian pigeons. Great Britain, and then the United States, tardily recognizing the great worth of birds that could carry messages when all other means failed, trained pigeons for military service. In World War II, private breeders supplied about two hundred thousand homing pigeons to the British services, and fifty thousand were reared by the U.S. Army. Released from aircraft that had crash-landed at sea, these birds bore SOS messages that saved many lives. Pigeons were also carried by trawlers and minesweepers that still lacked radios. Nearly seventeen thousand birds were parachuted to Resistance fighters in territories occupied by the Germans, and two thousand brought back reports from them.

In World War I, many pigeons won renown for heroic service on the battlefront. Among the most celebrated was Cher Ami, a blue-check male of British stock. In October 1918, a New York battalion of the American army had advanced so far into enemy territory that it was wholly surrounded by the foe. Every effort to report its isolation and desperate need of food and support failed until it had recourse to its carrier pigeons. The first few to be released fell, mortally wounded by an inferno of shells and shrapnel. The last hope was Cher Ami, who had already delivered twelve vital communications in the Verdun area. With a message giving the "Lost Battalion's" location attached to a leg, the winged courier rose through a barrage of bullets, circled, and turned toward his home loft at Rampont, when fragments of bursting shrapnel severed a leg and punctured his breast. Undaunted, the bird continued his flight, and in half an hour reached his destination, twenty-five miles (40 km) distant, bearing, on the leg that still hung from a few shreds of tissue, information that saved the battalion.

Equally distinguished, although less widely acclaimed, were the achievements of The Mocker, a red male with white markings. During the American advance into Alsace-Lorraine, he carried many vital messages. When heavy German artillery stopped the American troops, he bore from an observation post a report that enabled the American artillery to locate and silence the enemy's guns, opening the way for the capture of Beaumont. While he was carrying this message, shrapnel destroyed his left eye and gashed the top of his head, all without halting his flight. For his outstanding contributions to the Allied cause, he was awarded the American Distinguished Service Cross and the French *Croix de Guerre.* Although scarred and half-blind, the Mocker long survived the conflict, to die in 1937 at the age of twenty-one years.

A third celebrated pigeon of World War I was a black male, hatched in France and named President Wilson. He served with the tank corps and from airplanes and was officially credited with saving the lives of many American soldiers. On November 5, 1918, certain units in the Verdun sector, separated from the rest of the army, could not communicate with headquarters until they released the black pigeon. With a wounded breast and one leg shot away, through fog and heavy machine gun and artillery fire, he dauntlessly flew onward to his loft with the message attached to his remaining leg. After the war, the veteran was cared for at Fort Monmouth, New Jersey, until his death in 1929.

These homing pigeons, celebrated as heroes, were not greatly different from countless other pigeons—and birds of many other kinds—who, while migrating, pass through a barrage of huntsmen's bird shot. Despite severe wounds and pellets of lead embedded in their flesh, they continue bravely onward toward their destinations, until their forces fail or ebbing strength makes them easy victims of predators. With no grateful army to nurse them in an avian veterans' home until they die of old age, these casualties of a cruel "sport" fall unnoticed and expire unremembered.

The homing pigeons that served in the world wars were supplied by fanciers who raced their own birds. Pigeons intended for racing are born or bought while young and raised in lofts to which they become attached as their permanent homes. After they have nested there, their attachment becomes much stronger. By constant care, the fancier accustoms the birds to being handled. Every day, they are released to fly about the vicinity, learning to accompany their companions in compact flocks. They are trained to reenter the loft on hearing a particular sound, such as that of dry grain rattling in a tin, perhaps to the accompaniment of a distinctive whistle or call. They are fed only after their entry. If they are reluctant to enter the loft, they may be penalized by the withholding of food until they do better. After this preliminary education, selected individuals are trained for participation in a contest by being released at increasingly distant points, always in the same direction, from which they learn to fly home. The first release may be only 0.5 mile from the loft, to be followed by tosses into the air at 1 mile (1.6 km), then 2, next 3, with the intervals between successive tosses increasing up to a maximum of possibly 250 miles (400 km).

When a pigeon has been thoroughly trained to fly in a certain direction, and has become acquainted with the homeward route and the landscape around its loft, it may, on the day of the race, be released, along with many other birds from other lofts, much farther away, in wholly strange country, perhaps two or three times as far from home as its training took it. Obviously, it must now direct its course by some method of orientation that owes nothing to familiar landmarks. It must also be able to overcome pigeons' strong tendency to fly in flocks and thus must break away from slower birds and, when close to home, from those whose destination is not the same. Although pigeon-racers emphasize directional training, an experienced bird can find its way home from a point hundreds of miles away in the direction opposite to that in which it was trained.

For service in warfare, the pigeon's training is basically the same as that for racing, with one important difference. To be of help as an army advances or retreats, the bird must learn to return to a mobile rather than a stationary loft, one light enough to be transported by manpower or by a trailer. Young birds, preferably about five weeks old, are settled in it. In late afternoon or evening, they are released to fly around for only ten or fifteen minutes, taking exercise and becoming familiar with the surroundings, before they are called back for food. After they know their home and its location, they are tossed into the air at progressively longer distances from home, which widens their familiarity with the terrain. At intervals of a few days to a week, the loft is moved forward in the same direction, up to twenty miles (32 km) or more.

A major objective of this training is to accustom the birds to return to the loft, where they find food, water, salt, grit, and protection, wherever this refuge is situated. It is not easy for them to dissociate the loft from the landmarks that have guided them to it. Sometimes, after the loft has been advanced, confused pigeons circle briefly around it, then, ignoring the keeper's familiar voice and his rattling feed can, fly back to an earlier site and settle on the ground, seeking food on the exact spot where their loft formerly stood. They may visit in turn several such abandoned sites, until finally, if their loft has been moved many miles from its last position and they are in country new to them, they are irretrievably lost. If, in the removals of the loft made necessary by the chances of war, the pigeons returning to it after a toss meet flight routes that they have formerly followed, they may be diverted to these old familiar paths and fail to return to their loft in its latest situation, even after they have repeatedly homed to it.

A bird who can carry messages to and fro is more useful than one who can convey them only in one direction. Pigeons can be trained to serve as two-way couriers by being fed at a distance from the loft where they find everything else they need. They begin their education by being carried in a basket to a point in view of a stationary loft; they are fed in the basket and then permitted to fly home. Little by little, the basket is moved farther away; at each stage, the pigeons are called and fed, until it reaches its final destination. Here, any messages that they bear are removed from the containers attached to their legs, and replies are placed in the containers to be carried back to the home loft. The homeward flight is slightly slower than the outward flight because the birds are heavier with the grain they have eaten. Although this courier service must be centered on a stationary loft, the basket with food may be situated almost anywhere within a moderate distance from it.

Although many diurnal birds make long migratory journeys by night, pigeons prefer to travel by day. On long homing flights, as in races, they alight at dusk, on a ship if they are crossing water, to resume their journey next morning. However, migrating Wood Pigeons are known to continue their flights after nightfall, making it appear that homing pigeons, which belong to the same genus, might be trained to do the same. This training was tried by the British army, with moderate success. The birds were tossed, in twilight or after dark, at increasing distances from stationary lofts, where dim lights were screened from aerial observation. Carefully selected pigeons could be trained to return from points no more than twenty miles (32 km) from the loft, more slowly than in daylight. The French used night fliers in World War I, and the U.S. Army in World War II, but they proved to be of little military value.

That pigeons can be accustomed to fly through the night was demonstrated by fanciers in Hawaii, who established interisland races—for example, from Pluunene on Maui to Honolulu, a distance of about ninety-five miles (153 km), including two ocean crossings, of, respectively, eighteen and twenty-five miles (29 and 40 km). The birds traveled about as fast by night as they commonly do by day. The bold topography of the islands, with volcanoes rising high into the air, probably helped the birds direct their course. Only a minority of pigeons can home at night from distances of more than twelve miles (20 km). This ability appears to be inherited and to depend on nonvisual orientation, although it is difficult to exclude the possibility that the pigeons use landmarks or light patterns such as the sky glow from distant cities. Poor visibility hampers nocturnal homing.

People have raced horses, dogs, and other animals; but by far the longest races that depend wholly on muscular power are those of homing pigeons. Few sports require such prolonged, careful preparation as pigeon-racing. The birds must be bred, selected, and trained in directional flight by keepers in whom they gain confidence. The first pigeon race of over a hundred miles (160 km) was held in Belgium in 1818. After railroads offered swift transportation of homing birds to distant points of release, the races became longer and more frequent. In Britain alone, about a hundred thousand pigeon fanciers keep about two million birds in their lofts.

As the pigeons grow older, stronger, and more experienced, they are entered in increasingly long races. A course of about 100 miles (160 km) is considered suitable for juveniles a few months old, 300 miles (480 km) for yearlings, and 500 miles (800 km) for older birds. Birds based in Britain have been flown for 620 to 800 miles (1,000–1,300 km)—for example, from the Faeroe Islands in the north and from Spain in the south; but because of heavy losses over the sea, such demanding races are seldom flown. On the broad North American continent, races of up to 1,000 miles (1,600 km) are sometimes arranged, but no bird has reached home from this distance in one day. Only carefully selected pigeons, perhaps one in twenty, complete their preliminary training and return to their lofts from so far away.

Racing pigeons are sent on their long journeys with little food and water, so they may fly lighter, using stored fat as fuel, and be more eager to reach home and a meal. Separation from a nest with eggs or young, or perhaps from a newly won mate, may spur the racer to faster flight. There appears to be little agreement among fanciers about whether males or females are swifter.

In the races in which pigeons released at a single point reach separate destinations, it would be all too easy to cheat if elaborate precautions were not taken. At the time of release, an official places a rubber band with a code number on a leg of each competitor. The bird's owner or assistant, eagerly awaiting its return at the home loft, removes the band as soon as the bird arrives and drops it into a sealed box containing a clock that marks the hour and minute. The boxes are then carried to a central point and opened in the presence of judges who know the distance, over a great circle, from the point of release to the owner's loft, and who calculate the flier's speed in yards (or meters) per minute. The money deposited by entrants in the contest is used for prizes awarded to the owners of the swiftest birds. In good weather, homing speeds of twelve hundred yards per minute (41 miles, or 66 km, per hour) are frequent, and with a tail wind they may exceed two thousand yards per minute (68 miles, or 109 km, per hour). Indefatigable fliers may apparently continue in the air for sixteen hours a day, which is less time than some small birds are aloft on long overwater migratory flights. Pigeons who perform well in difficult races are sought by breeders, who offer high prices for them. Although much importance is attached to pedigree, the only trustworthy measure of a pigeon's ability is its performance in races.

Homing pigeons sometimes tire or lose their way, and seek hospitality in strange lofts, whose proprietors, if conscientious, may recognize their provenance by their bands and report them to their owners. Among the perils that confront racing pigeons are thunder, magnetic storms, and, above all, raptorial birds, which drive them from their courses and too frequently overtake them. Peregrine Falcons are a great menace. From two of their aeries on the Palisades along the Hudson River in New York, more than a hundred pigeons' bands were retrieved; one of them, recovered in 1932, was dated 1897. The bands lay among the remains of many birds, including Mourning Doves, woodpeckers, Blue Jays, Common Grackles, Rufous-sided Towhees, and a duck. Pigeons would more frequently return home if the skies were safer for them.

After serving humans as trustworthy messengers for thousands of years, homing pigeons are being given new tasks, such as carrying medical samples. On the northwest coast of France, a number of small hospitals send blood to a central laboratory for testing. With a test tube of blood, weighing about 1.4 ounces (40 g), fastened beneath its breast, a pigeon conveys the sample from Granville to the laboratory at Avranches, a distance of about sixteen miles (26 km), in twenty minutes, including the time for harnessing up, or more swiftly in a favoring westerly wind. This service is suspended only during the three-month hunting season in the autumn, when the risk of losing the trained pigeons to gunners is too great. The blood is then delivered in automobiles, which increases the cost of carriage by as much as forty-six dollars a day.

Although well-nourished pigeons can endure low temperatures, their feet are likely to freeze. For emergency service in polar regions, the Messenger Pigeon Project in Montana has been trying to develop a cold-tolerant strain of homing pigeons with well-feathered legs and toes. Although most pigeons have bare legs, some strains of the Rock Dove grow feathers on these limbs. By selection, the density of this covering might be increased.

Among the few other birds that have been employed as messengers are swallows, which, according to Pliny the Elder, were painted to carry home the colors of winning racehorses. In the Pacific, frigate birds were long used by Polynesians for interisland communication. Today, pigeons, as well as migratory birds, sometimes carry quite different messages. From tiny radios attached to their backs, they send signals that enable researchers to follow them, by day or night, in light aircraft. The signals yield fascinating information about the length and speed of the birds' flights.

11 How Pigeons Find Their Way

The training of homing pigeons is designed to familiarize them with territory increasingly distant from the home loft, and usually in one direction. Within the familiar region, they might direct their flight largely, it not wholly, by visual clues. However, researchers have learned that pigeons with eyes covered by frosted contact lenses, which make them unable to distinguish the forms of objects more than a few yards away, can find their way from distant points to within a mile or two (2 or 3 km) of home, evidently by other means of orientation. When they are suddenly carried far beyond the limits of their training, as in a long race, they find no well-known features of the terrain to guide them. They must determine the direction of home by different means. If their training flights have been consistently eastward, they may have a strong inclination to fly in this direction. But how do they orient themselves?

How pigeons find their way is one aspect of the wider problem of avian orientation and migration, and is most profitably considered in relation to this broader subject. A vast amount of research, in which docile, readily available homing pigeons have played a major role, has revealed that birds guide themselves through the air by diverse means. Of these methods, the most obvious and readily understood by us has been called "piloting," or guidance by easily detectable signs, which are usually visible landmarks but may be sounds, such as the roar of surf on a seacoast or the brawling of a mountain torrent, or even an odor associated with a definite locality. Some birds migrate along "leading lines": the seacoast, major rivers, mountain ranges, or other prominent geographic features that they can follow toward their destination, by day or on nights not too heavily clouded. Such clearly marked routes are available for only a fraction of the vast host of migrating birds, and, for many of them, for only part of the way.

Many birds use means of orientation more uniformly distributed over Earth. Beginning in late summer and continuing until winter's chill descends over much of the Northern Hemisphere, night skies are full of birds, from tiny warblers to big geese, winging southward to warmer lands. Nocturnal migration permits diurnal birds to forage while they interrupt their journey during the day; among passerines, diurnal travel is chosen chiefly by aerial flycatchers such as swallows and kingbirds, which can forage along the way.

When confined during the seasons of their migrations, small nocturnal migrants exhibit migratory restlessness, often designated by the German term *Zugunruhe;* they try to escape in the direction they would take if they were free. Indigo Buntings breed in the eastern and central parts of the

White-throated Pigeon *Columba vitiensis* Sexes alike. Philippines and New Guinea to Fiji and Samoa.

United States and winter mainly from southern Mexico through Central America to Panama. When Stephen Emlen tested them under the stars or in a planetarium with a setting normal for the season, they fluttered mostly southward in their circular cages in autumn and northward in spring. When the setting of the planetarium was reversed through 180 degrees, the directions taken by the buntings also tended to be reversed: northward in autumn and southward in spring. When the stars in the planetarium's dome were extinguished, the birds' directions were random.

These birds orient themselves by the Pole Star, which they recognize by its stationary position amid rotating constellations. When Emlen made the planetarium's sky circulate around Betelgeuse, the brightest star in the constellation Orion, juvenile Indigo Buntings, hand-reared with no opportunity to view the night sky, guided themselves by this star as though it were Polaris. Emlen demonstrated that young buntings lack an innate map of the heavens but learn to recognize the Pole Star (or a planetarium's substitute for it) by observation, probably of the constellations' changing positions rather than of their slow movements. As the date of their migrations approaches, birds that usually roost amid foliage must at least occasionally perch in the open, where they can see the stars. Birds were probably Earth's first star-gazers, watching the movements of the constellations ages before Chaldean and Babylonian astrologers paid attention to them. One wonders how migrants guide themselves when they fly so far southward that Polaris sinks below the horizon.

Many birds use the Sun for orientation, compensating for its changing positions in the sky with the passing hours by their internal clock, or circadian rhythm. The setting of this clock can be changed by keeping them in light-tight rooms, with artificial illumination out of step with daylight, until the birds are attuned to this altered schedule. Since pigeons travel mainly by day, they are much more likely to depend on the Sun than on the stars for guidance. Many experiments have been made to learn just how they use it. If their room is lighted from midnight to midday, their noon will correspond roughly to dawn outside; if the lights are turned on from noon until midnight, their midday will correspond to evening. Experiments have shown that when pigeons are clock-shifted by six hours and are released at a distance from home, they take directions roughly ninety degrees different from those of control birds on normal time—clockwise if their day has been retarded, counterclockwise if it has been advanced.

An objection to such experiments is that the birds in the artificially lighted room are deprived of the opportunity to see the Sun for so long a time that they forget where it was when they were enclosed; hence, they are unable to judge by its position at the time of the test how far or in what direction they have been carried. To meet this objection, Judith Alexander and William Keeton drove their experimental pigeons into an outside aviary during the morning, when their period of artificial illumination, from midnight to noon, coincided with daylight. Thus, the birds enjoyed abundant opportunity to observe the Sun without altering their daily period of illumination. When they were released twenty-one miles (33.5 km) east of their home loft, the control birds flew approximately westward and the time-shifted ones flew mainly toward the south. When they were tossed forty-six miles (73.5 km) north of home, the controls flew toward the south and the experimental birds generally eastward. These pigeons behaved much as they would have if they had been given no opportunity to see the Sun while their internal clocks were being reset. As in nearly all such tests, there was considerable scatter in the vanishing directions of both the experimental pigeons and the controls; but the mean direction of the time-shifted birds was eighty-three degrees counterclockwise from that of the control birds, which differs little from the expected ninety degrees.

Most of the pigeons on normal time arrived at their loft on the day of release; many of the clock-shifted birds failed to return home. If the pigeons whose clock had been advanced six hours had used the Sun to indicate their geographical location, they would have flown eastward from both points of release, for the Sun, far less advanced in its course across the sky than it would have been at home according to their altered time, was where they would have found it if they had been transported ninety degrees of longitude, or six time zones, to the west. Apparently, they were using the Sun only as a compass, orienting themselves by its azimuth, or horizontal direction, uninfluenced by its altitude in the sky.

The ability of birds to orient themselves by Earth's magnetic field, long suspected, was first demonstrated in the 1960s by Friedrich Merkel and his co-workers, who experimented with European Robins exposed to magnetic fields of varying orientation and intensity in steel chambers. Tiny crystals of magnetite (an oxide of iron) have been found in the heads of certain birds, where they may act as little lodestones pressing against ultrasensitive tissues. Magnets attached to pigeons flying under an overcast sky disorient them; brass bars of equal weight fastened to control birds do not. Pigeons can be turned about by electric coils, mounted on their heads, that change the angle of their magnetic field. By artificially shifting the field to which confined European Robins (short-distance migrants) were exposed, Verner Bingman altered the directions of their movements. When robins whose orientation had been shifted in this manner were tested under the stars in a vertical magnetic field that provided no directional information, their migratory movements were in the direction determined by the horizontal field to which they had previously been exposed. This and similar experiments suggest that Earth's magnetic field may be the birds' primary directional guide, to which others, such as the stars or the Sun, are calibrated.

The latest and most discussed hypothesis about the orientation of pigeons is that olfactory clues guide them, at least through a region within about sixty miles (100 km) from home. Scent-laden winds carry to their loft diverse odors from different directions, by the aid of which they develop an "odorous map," or mosaic of smells, of the surrounding territory. They associate one scent with a certain direction, a different odor with another; and when they are released at a moderate distance, they supposedly use this information to direct them home. In Italy, where the idea originated, a diversity of experiments have been made to support it, chiefly by Floriano Papi and his co-workers. In many tests, the pigeons' olfactory nerves were severed before they were released. To avoid the traumatic effects this operation might have, the researchers inserted tiny plastic tubes into the birds' nostrils in a way that permitted the birds to breathe without smelling. In other tests, a local anesthetic made them temporarily insensitive to odors.

A different approach was to rear the birds in lofts through which winds from only one direction would pass, or to deflect the breezes by screens so they reached the birds from a false direction. These are only a few of the ingenious methods that were devised to shift the orientation of the birds or to prevent the formation or use of odorous maps. Pigeons so treated were released at the same time and place as control pigeons, and their homing performances were compared. Frequently, as expected, the experimental birds did not do as well as the controls, presumably because they could not detect odors or were led astray by their previous exposure to winds of altered directions. Nevertheless, some of the experimental birds found their way home.

Although birds appear to be more sensitive to odors than is commonly believed, many objections have been raised against this hypothesis. The supposedly effective odors have not been identified, although they are thought to be derived from vegetation and to be molecules rather than particles or aerosols. When the Italian experiments were repeated in the vicinity of Ithaca, New York, the results were less convincing. The differences might be due to the diverse nature of the two regions, different strains of birds, or different ways of raising and training them. Meteorologists' studies of the diffusion of scents in the atmosphere have cast doubt on the feasibility of olfactory orientation. Nevertheless, after a careful appraisal of all the evidence, Klaus Schmidt-Koenig conceded that airborne scents, in conjunction with other information, might help pigeons, especially inexperienced young birds, to home over short distances. This other information might be provided by Earth's magnetic field.

Experiments and field observations have yielded abundant evidence that individual birds are equipped with several means of orientation, so that in situations where one navigational clue is unavailable, they can rely on another. In view of the vital importance to many kinds of birds of finding their way over great distances, it is not surprising that they have evolved complementary methods of guiding themselves. This diversity of resources for orientation has complicated and confused their study, making it difficult to investigate any one of them to the exclusion of the others.

In pigeon races for which the birds have been thoroughly trained, they need only fly in the predetermined direction to reach a fairly wide area familiar to them, after which they can find their way home without difficulty. A crosswind may blow them off course, or they may lose their way and find themselves in strange country, not in the predetermined direction from their base. Nevertheless, some of these straying pigeons do eventually reach their lofts. If they do not find them by random wandering, which would seem rarely to bring them home from afar, they must use some method more sophisticated than the sailor's "distance and bearing" navigation that would have taken them to their destination if nothing had gone amiss.

However it might be with pigeons, there is abundant evidence from banded birds that long-distance migrants commute yearly between two definite points, such as a nest site in the north and a garden in the south where they winter, separated by thousands of miles. For such precise navigation, the most accurate compass, whether it be magnetic, the Sun, or the stars, does not suffice. As every airplane pilot must know, a map, or something that corresponds to it, is indispensable. One must know one's location on the surface of the globe, the location of the destination, and the direction of the second point from the first.

As we have seen, birds have no lack of means for determining direction. What they use for a map or grid, corresponding to the lines of longitude and latitude in an atlas, is less clear. In the history of exploration, the latitude of newly discovered lands was readily determined by measuring the angle of the Sun's altitude above the horizon at noon, or by the stars, but longitude presented greater difficulties in the absence of chronometers to indicate the difference between the explorer's local time and that at a fixed point of reference, such as Greenwich. Likewise, birds might learn whether they are north or south of their known destination by observing whether the Sun at noon is lower or higher than they remembered it to be, allowing for seasonal differences. By means of their internal clock, they might even gauge their longitudinal separation from their destination. G. V. T. Matthews postulated that even brief observation of the Sun's course through the sky might enable a bird, by extrapolation, to determine its elevation at noon and its own geographical position in relation to its goal. This hypothesis attributes to birds such great powers of observation and rapid deduction that, in the absence of unequivocal experimental evidence, it has been rejected.

The strength, horizontal direction, and tilt, or vertical inclination, of Earth's magnetic field may offer clues to geographical location. It has been suggested that variations in the Coriolis force that acts on a body moving through the gravitational field of a revolving planet may be detected by a bird, although this force is too weak to be felt directly by man. Lines of equal Coriolis force intersect with those of the magnetic field to form a grid that the birds might use for navigation. The experiments designed to demonstrate this use were inconclusive.

The quest for deeper understanding of the homing ability of pigeons has prompted innumerable experiments. Do the birds register every change of direction and acceleration on the outward journey and integrate this information to find their way home? Investigations of the possibility of inertial navigation, improbable as it seems, have involved transporting the birds in drums revolving with such complex, irregular movements that the birds have become seasick, anesthetizing them during the outward journey, and surgically severing the circular canal of the inner ear (the organ of balance), all without preventing their return to their lofts. Testing this idea and many other guesses about how birds orient and navigate has resulted in the mutilation of many unfortunate pigeons. In ordinary practice, the birds are carried to the point of release in closed containers, so that they cannot see where they are going. The evidence points to the conclusion that on their release, and on the homeward journey, they obtain the information they need to direct their course.

Do pigeons fly off in the proper direction, and find their way home, better when released singly or in small flocks? To answer this question and others about their homing performance, researchers have followed them in airplanes, or watched them with binoculars from the point of release as long as they could be kept in view, noting with a compass the direction of their flight, then marking their vanishing points on the circumference of a circle. The use of binoculars is much less costly than the use of planes and on the whole has proved more satisfactory, for the pigeons might be disturbed by the noisy aircraft flying in pursuit of them like a wide-winged raptor. As in many experiments designed to study homing pigeons, results have been contradictory, probably in consequence of differences in the homing ability of the birds or in their training. Some tests have indicated that they perform better in flocks. However, in a carefully planned experiment, William Keeton selected birds of similar age and training and released, alternately, single pigeons and groups of four. He found no significant difference in the orientation at the point of release of solitary birds and quartets. In most tests, all arrived home in comparable times; but in one, the flocks were faster. If one bird in a flock is a superior homing pigeon and its companions follow its lead, the flock should perform better than the average individual. In the absence of such a guide, we should expect no consistent difference in the performances of singles and groups.

Despite many ingenious theories and a vast amount of experimentation and field observation, we are far from understanding what migrating or homing birds use for a map or grid. As Kenneth Able concluded, "The ability of true navigation found in birds remains one of the most enduring mysteries in biology." When we contemplate navigational feats that baffle our comprehension, performed by birds that lack the charts, instruments, and special training that humans need for comparable accomplishments, we are filled with profound respect for their abilities while we continue to wonder what goes on inside their little heads.

12 Darwin's Pigeons

When, as naturalist on the surveying ship H.M.S. *Beagle,* Charles Darwin visited the Galápagos Archipelago in 1835, he was impressed by the way closely related species of animals varied from island to island. They started a train of thought on the mutability of organisms which, years later, ripened in *The Origin of Species.* The most numerous of the land birds of those isolated volcanic islands, six hundred miles (nearly 1,000 km) west of South America, are fourteen species of finches which, apparently from a common ancestor derived from the continent, have become modified for diverse modes of life. Although the visiting naturalist seemed to pay more attention to the mockingbirds and tortoises, he did not forget the finches, which have become familiarly known as "Darwin's finches."

After becoming engrossed in the problem of the origin of species, Darwin, now far from the Galápagos, which he never revisited, turned his attention to domestic pigeons, readily available in England, where he spent the rest of his life. Believing that it is always best to study some special group, after deliberation, he maintained every breed of pigeons he could purchase or otherwise obtain. Through correspondence, he procured many of them from distant lands, especially India and Persia (now Iran). He joined two London pigeon clubs, and he read or had translated for him some of the many treatises on these birds which for centuries had been written in different languages. In *The Origin of Species,* pigeons receive far more attention than Galápagos finches, which are hardly mentioned by name; and they are the subjects of two long chapters in *The Variation of Animals and Plants under Domestication.* Not without reason did George Bernard Shaw, in the preface to *Back to Methuselah,* describe Darwin as "an intelligent and industrious pigeon fancier," although, of course, he was much more than that—a great philosophical naturalist with exceptionally wide interests. Certainly, it is as proper to speak of Darwin's pigeons as of Darwin's finches. Of all the contributions the study of pigeons has made to science, none is more important than its influence on Darwin's thought about evolution, which has so greatly changed our views of our world and ourselves.

I venture to assert that, until the recent upsurge in recreational bird-watching, pigeons formed the strongest affective bond between man and the feathered world. During the millenia that fanciers in many countries have bred pigeons for their beauty, bizarreness, or other qualities, the birds have diverged so greatly from their ancestor, the wild Rock Dove (or Rock-Pigeon, as Darwin called it), that, as he remarked, if they were free birds rather than domestic varieties, ornithologists would classify them in different species or even genera. Nevertheless, he believed that the evidence that all these astonishingly different races had descended from one known source was far stronger than in the case of any other animal that had been domesticated in ancient times.

Darwin estimated the number of distinguishable kinds of Domestic Pigeons as over 150, but this number included many that differed only slightly from others. He classified them in four main groups, and all but the first in several subgroups. His first group comprised a single race, the Pouters, of which the improved English subrace appeared to be the most distinct of all domesticated pigeons.

Pigeons and doves commonly puff out their chests to give resonance to their coos. Breeders have seized on this widespread habit to create, by generations of selection, a pigeon that can inflate its chest, even to just below its mouth, to truly astonishing size, thanks to the extraordinary diameter of its esophagus. One of Darwin's Pouters could puff itself up until its beak was almost wholly hidden. Males, especially when excited, pout more than females and seem to delight in the performance. Sometimes, a fancier, picking up a pigeon slow to inflate itself, puts the bird's beak in his own mouth and blows him up like a balloon, whereupon the Pouter, "puffed up with wind and pride, struts about, retaining his magnificent size as long as he can" (Darwin 1900). About nineteen inches (48 cm) long, the English Pouter is larger than the ancestral Rock Dove. Its legs and toes are often feathered. In flight, it often strikes its wings together above its back, making a loud clapping noise.

Apparently, Darwin did not know of a related race called the Modenas, from the Italian city in which it originated in medieval times, if not earlier. These birds were selected for their ability to entice individuals from other lofts to join their flocks and follow them back to the decoys' home, where they were held for ransom or, in a less neighborly spirit, killed. This game was reported to be widely practiced in Mediterranean countries. In the United States, Modenas are valued for their robust appearance and many attractive color patterns.

In group II, Darwin's second race included the English Carrier, a large, long-necked, generally dark-colored pigeon with an exceptionally long bill and tongue. The ring of bare skin around the eye, smooth in most pigeons, is often unusually wide and corrugated or wattled, sometimes excessively, in the English Carrier, and the cere at the base of the bill is hypertrophied into a fleshy growth that extends over the nostrils and to the lower mandible. In chapter 1, we noticed a similar enlargement of the cere in a few wild pigeons unrelated to the Rock Dove. On pigeons used for racing and carrying messages, the wattles tend to be less exaggerated. A subrace known as the English Dragon was smaller than the improved English Carrier, with less wattle around the eyes and over the nostrils, none on the lower mandible.

The exceptionally large pigeons of Darwin's third race, facetiously called Runts, also known as Scanderoons, have wattled skins around the eyes and over the nostrils. When the birds are not very big, they grade into Carriers, making the two races difficult to separate. Scanderoons have short, narrow, uptilted tails and extremely short wings, but long necks and bills.

The fourth race consists of the Barbs, large pigeons with broad heads and short, thick bills like those of Bullfinches. Their plumage is generally dark

English Pouter, a variety of Domestic Pigeon *Columba livia.* (From Darwin 1900.)

and plain; their legs are short and stout. Their eyes are encircled by broad rings of naked skin, which is sometimes so strongly carunculated that the poor birds can hardly see to pick up their food. This feature allies the Barb to the Carrier. However, the swollen skin above the nostrils is not corrugated, except slightly in old birds.

In group III, the Fantails, Darwin's fifth race, are distinguished by their spreading uptilted tails, which may be composed of from twelve feathers (the normal number in the genus *Columba*) to thirty-four, and occasionally as many as forty-two, arranged in a double row. The erect, fanlike carriage of the tail, which gives these pigeons a distinct and elegant appearance, is more valued by fanciers than is the number of rectrices that compose it. To support this tail, the uropygium has one or two extra vertebrae, but the oil gland at its base is aborted. Fantails are broad-breasted birds of diverse colors; they walk stiffly on short legs, with their thin necks bent so far backward that their heads touch and ruffle their tails. Probably because of the unusual curvature of their necks, they bob their heads convulsively instead of smoothly, as do other pigeons, when they walk. They habitually tremble so much that they were formerly known as Shakers. Incommoded by their tails, they fly clumsily in a wind.

The Turbits and Owls of Darwin's sixth race are small, pretty pigeons with short, vertically thick bills. Their distinguishing feature is a frill of irregularly spreading feathers along the front of the neck and breast. As though to call attention to this adornment, they continually shake it by rapidly inflating and contracting the dilated upper end of the esophagus. The Turbit wears a crest on its head, which the Owl lacks.

The Tumblers, Darwin's seventh race, are fantastic birds that turn somersaults, on the ground or in the air. If an Indian Ground Tumbler is gently shaken and placed on the ground, it promptly begins to tumble head over heels, continuing this amazing performance until it is picked up and calmed, which is usually accomplished by one's blowing gently in its face, much as one might do to arouse a hypnotized person. If not rescued from its falling spell, the unfortunate pigeon may continue until it expires from exhaustion. Some Tumblers begin to tumble when touched on the neck with a stick or cane. English Tumblers begin to perform almost as soon as they can fly, and they improve in the following months. When older, they tumble in the air, turning twenty to thirty somersaults in a minute, occasionally as many as forty. Some individuals fly like a revolving wheel, continuing until they strike the ground, sometimes killing themselves. Others, called House Tumblers, turn two or three somersaults while flying across their loft; unlike the Indian variety, they do not need the stimulus of a shake to set them tumbling.

Another subrace, the Short-faced Tumblers, were extolled by Darwin as "marvellous birds . . . the glory and pride of many fanciers." Their heads are nearly globular and sometimes bald, the foreheads high. Their bills are short, sharp, and conical, with reduced ceres. On very short legs, they walk erect, breasts protruding, wings trailing. They seldom tumble. They weigh less than half as much as a wild Rock Dove, about one-fifth as much as a large Runt. These Tumblers, afflicted by some nervous derangement that would promptly cause their extinction in the wild, have been preserved for centuries in the aviaries of their admirers.

The eighth race is the Indian Frill-back, a rather small, short-billed pigeon with all its feathers reversed or curled backward, much as in a freakish variety of domestic chickens. Next comes the Jacobin, readily recognized by its hood, formed by the elongated, forwardly curled feathers of its nape, which almost enclose its head and meet in front of its neck. It reminds me of the black hoods of male umbrella birds of tropical American forests but is evidently more enveloping. The Jacobin's wings and tail are also greatly elongated, but its beak is short.

The races and subraces of group IV differ little in structure, especially in that of the bill, from wild Rock Doves. Among those in Darwin's aviaries, the most distinct was the large Trumpeter, whose voice was wholly unlike that of any other pigeon he knew. For several minutes it continued rapidly to repeat its coo. A tuft of elongated feathers curled forward over the base of its bill, as in no other of Darwin's pigeons. Its small feet were so densely feathered that they almost appeared to be little wings. Among the so-called toy pigeons was the Laugher, scarcely different in structure from the Rock Dove but smaller, with a most peculiar voice. The Common Frill-back was larger than the Rock Dove and wore feathers that curled upward or backward, above all on its wing coverts. Nuns were small, elegant pigeons, with white bodies and contrastingly colored heads, primary wing feathers, tails, and tail coverts, all either black or red. Young Nuns had leaden black legs and toes, which appeared remarkable to Darwin, as the leg color of adults and immatures differs little in most breeds. Another white-bodied subrace was called Spots, for the spot on its forehead, of the same color as its tail and tail coverts. Swallows were large, slender pigeons, with small legs and feet, colored heads and wings contrasting with white bodies, and a peculiar manner of flying.

Darwin knew of other breeds of pigeons which he did not describe. Some of the races he kept were already centuries old, perhaps dating from ancient Rome, when pigeons were cherished by fanciers, as in our own day. How many of the breeds known to Darwin are extant, and how many new ones have been developed since his time, I cannot tell, as I am not acquainted with any pigeon fancier and the publications of aviculturists are unavailable here. Ornithologists who seriously study free pigeons sometimes scorn as "monstrosities" these cosseted breeds, unable to maintain themselves without human pampering. Oddities many of them certainly are; but car-ing for them has for many people been the joy of a lifetime, a wholesome relaxation on stressful days, a solace in sorrow and adversity. Often, fanciers breed their pigeons according to an ideal of beauty or perfection of the race they most admire, which might differ greatly from that preferred by another fancier. And some of the domestic breeds are far from being monstrosities. The pure white Maltese and rich maroon Carneaux that I loved when I was a boy were normal pigeons that differed chiefly in color from the common pigeons who lived with them.

The pigeons that Shaw's "intelligent and industrious pigeon fancier" raised and knew so well provided what we might call his showcase for the mutability of species, his irrefutable evidence that animals change their forms and colors over the generations. He admitted that when at first he entered his aviaries and watched such birds as Pouters, Carriers, Barbs, Fantails, Short-faced Tumblers, and others, he could not persuade himself that all had descended from the same wild stock, and that man had consequently, in one sense, created these remarkable modifications. However, after carefully weighing all the evidence, he had no doubt that this was true.

Darwin refuted the belief that the chief domestic races of pigeons were descended from at least eight, nine, or possibly a dozen species—the smallest number that might, by hybridization, yield the characteristic differences among these races—by six strong arguments. The first is the improbability that so many species should still exist somewhere, unknown to ornithologists, or should have become extinct within historical times, although man had done little to exterminate the wild Rock Doves. The second is the improbability that in former times man had thoroughly domesticated so many species, with no reduction of fertility in confinement. The third is the fact that these supposed species have nowhere become feral. The fourth is the improbability that man should have, intentionally or by chance, chosen for domestication species with extremely abnormal characters that remain so highly variable. The fifth is the fact that all these races, so different in many important structural features, produce perfectly fertile hybrids, whereas all the hybrids that result from crossing even closely allied species of pigeons are sterile. The sixth is the tendency of all the domestic races, both when purebred and when crossed, to revert in numerous minute details to the coloring of the wild Rock Dove and to vary among themselves in a similar fashion. Thus, the offspring of a cross between two distinct domestic races, neither of which shows a trace of blue, and probably has not done so for many generations, are sometimes blue like the ancestral Rock Dove, with perhaps also the black wing bars characteristic of that bird. Finally, all these races, so different in appearance, are so similar in other details of structure, as well as in habits and voice.

Darwin attributed the amazing diversity of Domestic Pigeons to prolonged selection by man. He distinguished two kinds of artificial selection.

Methodical selection is practiced by a breeder with a definite goal, such as the preservation of a character that has already appeared or the improvement, in quality or yield, of some cultivated plant. Unconscious selection results, as Darwin remarked, from man's rivalry, his effort to outdo neighbors. Fanciers are never content with pigeons that fall short of the prevailing standard of their preferred race, and are never more pleased than when they can intensify its peculiar character, as by producing a shorter beak on a short-billed variety or a longer beak on long-billed stock. The unforeseen result of this effort may be a distinct race.

Preference for exaggerated, or supernormal, characters is not confined to man. A Ringed Plover prefers eggs more heavily spotted than its own. An Oystercatcher neglects its own eggs to try vainly to incubate one too big for it to sit on, as though infatuated by its size. Such preference for intensified characters by female birds choosing mates may be largely responsible for the extravagantly beautiful plumage of the males of certain birds of paradise and other birds, especially those whose emancipation from nest attendance leaves them free to become conspicuous with feathers or wattles that would reduce their efficiency as attendant parents. Among birds of which both parents share domestic tasks, perhaps we might attribute the magnificent crests of crowned pigeons and the lovely colors of fruit doves to the same psychic trait that in man has led to the development of very different races of domestic pigeons, each beautiful in the eyes of its fanciers—that of preferring an intensified expression of a conspicuous character, which acts like a supernormal stimulus. It is sobering to reflect on how much we have in common with animals that we deem inferior to ourselves.

Darwin's thought passed from the striking effects of selection as practiced by man on his domestic animals and plants to the importance of selection in nature's wider realm. He attributed to it a major role in the evolution of organisms, as is evident from the full title of his major work: *On the Origin of Species by Means of Natural Selection, or the Preservation of Favoured Races in the Struggle for Life.* But natural selection and artificial selection, in spite of certain similarities, contrast sharply in their methods and their results. Competent breeders of animals or plants give preferential treatment to the individuals that show improvement in the characters for which they are selected. The breeders feed or cultivate them more carefully and protect them more vigilantly, without necessarily penalizing or destroying the less favored individuals. Nature does nothing of the sort; it eliminates, by predation, disease, starvation, or other harsh means, the less well adapted individuals, while taking no special care of the better adapted. In this sense, artificial selection is positive; natural selection, negative.

On the other hand, artificial selection, undertaken for man's benefit and subject to his caprice, intensifies some character at the expense of the overall fitness of the animal or plant on which it operates. Food plants may be made more productive at the price of their hardiness, and flowers larger at the expense of their fertility; animals (such as some of the fancy breeds of pigeons) may be so strangely altered from their ancestral forms that they are incapable of surviving without human pampering. Acting on a wider spectrum of characters than artificial selection commonly takes into account, natural selection improves a species' adaptation to its environment indirectly by removing from the gene pool the less well adapted genes. Natural selection is essentially natural elimination, whereas artificial selection takes a more direct approach to its goals, however limited or dictated by human whims they may be.

White-faced Pigeon *Turacoena manadensis*
Also called White-faced Cuckoo-Dove.
Sexes alike. Sulawesi (Celebes),
Sula Islands, and neighboring islands.

Bibliography

Chapter 1 The Pigeon Family
(General references, distribution, anatomy, classification, migration, longevity)

Bent, A. C. 1932. *Life Histories of North American Gallinaceous Birds.* U.S. National Museum Bulletin no. 162.

Campbell, B., and E. Lack, eds. 1985. *A Dictionary of Birds.* Calton, England: T. and A. D. Poyser.

Cottam, C., and J. B. Trefethen, eds. 1968. *Whitewings: The Life History, Status, and Management of the White-winged Dove.* Princeton, N.J.: D. Van Nostrand Co.

Frith, H. J. 1982. *Pigeons and Doves of Australia.* Adelaide: Rigby Publishers.

Goodwin, D. 1967. *Pigeons and Doves of the World* (with addendum to 2d ed., 1977). London: British Museum (Natural History).

Murton, R. K. 1965. *The Wood Pigeon.* London: Collins.

Rowley, I. 1975. *Bird Life.* Sydney and London: Collins.

Terres, J. K. 1980. *The Audubon Society Encyclopedia of North American Birds.* New York: Alfred A. Knopf.

Thompson, D. R. 1950. Foot-freezing and arrested post-juvenal wing molt in the Mourning Dove. *Wilson Bulletin* 62:212–213.

Chapter 2 Eating and Drinking

Alcorn, S. M., S. E. McGregor, and G. Olin. 1961. Pollination of saguaro cactus by doves, nectar-feeding bats, and honeybees. *Science* 133:1594–1595.

Bent, A. C. 1932. (Full reference in list for chapter 1.)

Cottam, C., and J. B. Trefethen, eds. 1968. (Full reference in list for chapter 1.)

Fisher, C. D., E. Lindgren, and W. R. Dawson. 1972. Drinking patterns and behavior of Australian desert birds in relation to their ecology and abundance. *Condor* 74:111–136.

Goodwin, D. 1967. (Full reference in list for chapter 1.)

Grant, P. R., and K. T. Grant. 1979. Breeding and feeding ecology of the Galápagos Dove, *Condor* 81:397–403.

MacMillen, R. E., and C. H. Trost. 1966. Water economy and salt balance in White-winged and Inca doves. *Auk* 83:441–456.

Murton, R. K. 1965. (Full reference in list for chapter 1.)

Phelan, J. P. 1987. Some components of flocking behavior in the Rock Dove (*Columba livia*). *Journal of Field Ornithology* 58:135–143.

Pratt, T. K. 1984. Examples of tropical frugivores defending fruit-bearing plants. *Condor* 86:123–129.

Skutch, A. F. 1964. Life histories of Central American pigeons. *Wilson Bulletin* 76:211–247.
Skutch, A. F. 1983. *Birds of Tropical America.* Austin: University of Texas Press.
Webster, M. D., and M. H. Bernstein. 1987. Ventilated capsule measurements of cutaneous evaporation in Mourning Doves. *Condor* 89:863–868.
Wiley, J. W., and B. N. Wiley. 1979. *The Biology of the White-crowned Pigeon.* Wildlife Monographs 64.
Willoughby, E. J. 1966. Water requirements of the ground dove. *Condor* 68:243–248.

Chapter 3 Daily Life

Bastin, E. W. 1952. Flight speed of the Mourning Dove. *Wilson Bulletin* 64:47.
Bent, A. C. 1932. (Full reference in list for chapter 1.)
Brackbill, H. 1970. New light on the Mourning Dove. *Maryland Magazine* 2:8–10.
Bucher, E. H. 1970. *Consideraciones ecológicas sobre la paloma* Zenaida auriculata *como plaga en Córdoba.* Ministerio de Economía y Hacienda, Serie Ciencia y Técnica no. 1:1–11.
Cowan, J. B. 1952. Life history and productivity of a population of western Mourning Doves in California. *California Fish and Game* 38:505–521.
Darwin, C. 1859. *On the Origin of Species by Means of Natural Selection.* London: John Murray (and many later editions).
Dorst, J. 1957. The puya stands of the Peruvian high plateau as a bird habitat. *Ibis* 99:594–599.
Goodwin, D. 1967. (Full reference in list for chapter 1.)
Hauser, D. C. 1957. Some observations on sun-bathing in birds. *Wilson Bulletin* 69:78–90.
Johnston, R. F. 1960. Behavior of the Inca Dove. *Condor* 62:7–24.
MacMillen, R. E., and C. H. Trost. 1967. Nocturnal hypothermia in the Inca Dove *Scardafella inca. Comparative Biochemistry and Physiology* 23:243–253.
Marshall, J. T., Jr. 1949. The endemic avifauna of Saipan, Tinian, Guam, and Palau. *Condor* 51:200–221.
Meinertzhagen, R. 1955. The speed and altitude of bird flight. *Ibis* 97:81–117.
Murton, R. K. 1965. (Full reference in list for chapter 1.)
Petersen á Botni, N. F., and K. Williamson. 1949. Polymorphism and breeding of the Rock Dove in the Faeroe Islands. *Ibis* 91:17–23.
Robertson, P. B., and A. F. Schnapf. 1987. Pyramiding behavior in the Inca Dove: Adaptive aspects of day-night differences. *Condor* 89:185–187.
Skutch, A. F. 1964. (Full reference in list for chapter 2.)
Smythies, B. E. 1960. *The Birds of Borneo.* London: Oliver and Boyd.
Sooter, C. 1947. Flight speeds of some South Texas birds. *Wilson Bulletin* 59:174–175.
Wiley, J. W., and B. N. Wiley. 1979. (Full reference in list for chapter 2.)

Chapter 4 Voice and Courtship

Bent, A. C. 1932. (Full reference in list for chapter 1.)

Burley, N. 1981. Mate choice by multiple criteria in a monogamous species. *American Naturalist* 117: 515–528.
Fleay, D. 1961. The Gouras of New Guinea. *Animal Kingdom* 64:106–110.
Frith, H. J. 1977. Some display postures of Australian pigeons. *Ibis* 119:167–182.
Goodwin, D. 1966. The bowing display of pigeons in reference to phylogeny. *Auk* 83:117–123.
Goodwin, D. 1967. (Full reference in list for chapter 1.)
Johnston, R. F. 1960. (Full reference in list for chapter 3.)
Marshall, J. T., Jr. 1949. (Full reference in list for chapter 3.)
Murton, R. K. 1965. (Full reference in list for chapter 1.)
Murton, R. K., and A. J. Isaacson. 1962. The functional basis of some behaviour in the Woodpigeon *Columba palumbus*. *Ibis* 104:503–521.
Peeters, H. J. 1962. Nuptial behavior of the Band-tailed Pigeon in the San Francisco Bay area. *Condor* 64:445–470.
Saunders, D. C. 1951. Territorial songs of the White-winged Dove. *Wilson Bulletin* 63:330–332.

Chapters 5, 6, and 7 Nests and Eggs, Incubation, and The Young and Their Care

Anderson, A. H., and A. Anderson. 1948. Observations on the Inca Dove at Tucson, Arizona. *Condor* 50:152–154.
Bent, A. C. 1932. (Full reference in list for chapter 1.)
Bent, A. C. 1940. *Life Histories of North American Cuckoos, Goatsuckers, Hummingbirds, and their Allies.* U.S. National Museum Bulletin no. 176.
Blockstein, D. E. 1986. Nesting trios of Mourning Doves. *Wilson Bulletin* 98:309–311.
Bucher, E. H. and M. Nores. 1973. Alimentación de pichones de la paloma *Zenaida auriculata. Hornero* 11:209–216.
Bucher, E. H., and A. Orueta. 1977. Ecología de la reproducción de la paloma *Zenaida auriculata.* II: Epoca de cría, suceso y productividad en las colonias de nidificación de Córdoba. *Ecosur* 4:157–185.
Crome, F. H. J. 1975a. Notes on the breeding of the Purple-crowned Pigeon. *Emu* 75:172–174.
Crome, F. H. J. 1975b. Breeding, feeding and status of the Torres Strait Pigeon at Low Isles, North-eastern Queensland. *Emu* 75:189–198.
Fleay, D. 1961. (Full reference in list for chapter 4.)
Fraga, R. M. 1983. Conducta vocal y reproductiva de la Yeruti Común (*Leptotila verreauxi*) en Lobos, Buenos Aires, Argentina. *Hornero* 12:89–95.
Franks, E. C. 1967. The responses of incubating Ringed Turtle Doves (*Streptopelia risoria*) to manipulated egg temperatures. *Condor* 69:268–276.
Goforth, W. R. 1964. Male Mourning Dove rears young unaided. *Auk* 81:233.
Goodwin, D. 1947. Breeding-behaviour in Domestic Pigeons four weeks old. *Ibis* 89:656–658.
Goodwin, D. 1967. (Full reference in list for chapter 1.)
Hanson, H. C., and C. W. Kossack. 1963. *The Mourning Dove in Illinois.* Carbondale: Southern Illinois University Press.
Haverschmidt, F. 1953. Notes on the life history of *Columbigallina talpacoti* in Surinam. *Condor* 55:21–25.

Hitchcock, R. R., and R. E. Mirarchi. 1984a. Duration of dependence of wild fledgling Mourning Doves upon parental care. *Journal of Wildlife Management* 48:99–108.
Hitchcock, R. R., and R. E. Mirarchi. 1984b. Comparison between single parent and normal Mourning Dove nestlings during the post-fledging period. *Wilson Bulletin* 96:494–495.
Johnston, R. F. 1960. (Full reference in list for chapter 3.)
Luther, D. M. 1979a. An intensive study of parental behavior in the Mourning Dove. *Indiana Audubon Quarterly* 57:209–232.
Luther, D. M. 1979b. Behavior of Mourning Doves during a 133-day incubation period. *Indiana Audubon Quarterly* 57:232–234.
McClure, H. E. 1944. Nest survival over winter. *Auk* 61:384–389.
McClure, H. E. 1945. Reaction of the Mourning Dove to colored eggs. *Auk* 62:270–272.
Marchant, S. 1960. The breeding of some S. W. Ecuadorian birds. *Ibis* 102:349–382.
Maridon, B., and L. C. Holcomb. 1971. No evidence for incubation patch changes in Mourning Doves throughout incubation. *Condor* 73:374–375.
Matthews, L. H. 1939. Visual stimulation and ovulation in pigeons. *Proceedings of the Royal Society of London,* series B, 226:423–456.
Murton, R. K. 1965. (Full reference in list for chapter 1.)
Murton, R. K., E. H. Bucher, M. Nores, E. Gómez, and J. Reartes. 1974. The ecology of the Eared Dove (*Zenaida auriculata*) in Argentina. *Condor* 76:80–88.
Murton, R. K., and A. J. Isaacson. 1962. (Full reference in list for chapter 4.)
Neff, J. A. 1944. A protracted incubation period in the Mourning Dove. *Condor* 46:243.
Neff, J. A., and R. J. Niedrach. 1946. Nesting of the Band-tailed Pigeon in Colorado. *Condor* 48:72–74.
Nice, M. M. 1922–23. A study of the nesting of Mourning Doves. *Auk* 39:457–474; 40:37–58.
Peeters, H. J. 1962. (Full reference in list for chapter 4.)
Petersen á Botni, N. F., and K. Williamson. 1949. (Full reference in list for chapter 3.)
Peterson, A. T. 1986. Rock Doves nesting in trees. *Wilson Bulletin* 98:168–169.
Raney, E. C. 1939. Robin and Mourning Dove use the same nest. *Auk* 56:337–338.
Schooley, J. P., and O. Riddle. 1944. Effect of light upon time of oviposition in Ring-Doves. *Physiological Zoology* 16:187–193. Abstract in *Bird-Banding* 16:43 (1945).
Skutch, A. F. 1949. Life history of the Ruddy Quail-Dove. *Condor* 51:3–19.
Skutch, A. F. 1956. Life history of the Ruddy Ground Dove. *Condor* 58:188–205.
Skutch, A. F. 1959. Life history of the Blue Ground Dove. *Condor* 61:65–74.
Skutch, A. F. 1964. (Full reference in list for chapter 2.)
Skutch, A. F. 1976. *Parent Birds and Their Young.* Austin: University of Texas Press.
Skutch, A. F. 1981. *New Studies of Tropical American Birds.* Publication no. 19. Cambridge, Mass.: Nuttall Ornithological Club.
Skutch, A. F. 1983. (Full reference in list for chapter 2.)
Skutch, A. F. 1987. *Helpers at Birds' Nests: A Worldwide Survey of Cooperative Breeding and Related Behavior.* Iowa City: University of Iowa Press.
Van Someren, V. G. L. 1956. Days with birds: Studies of habits of some East African species. *Fieldiana Zoology* 38:1–520.

Weeks, H. P. 1980. Unusual egg deposition in Mourning Doves. *Wilson Bulletin* 92:258–260.
Westmoreland, D., and L. B. Best. 1986. (Full reference in list for chapter 8.)
Whitman, C. O. 1919. *The Behavior of Pigeons.* Publication 257. Washington, D.C.: Carnegie Institution. (Not seen by me.)
Wiley, J. W., and B. N. Wiley. 1979. (Full reference in list for chapter 2.)
Willoughby, E. J., and C. T. Krebs. 1986. Adaptability of parental behavior in the Mourning Dove. *Journal of Field Ornithology* 57:238–239.

Chapter 8 Rate of Reproduction

Bucher, E. H., and A. Orueta. 1977. (Full reference in list for chapters 5, 6, and 7.)
Burley, N. 1980. Clutch overlap and clutch size: Alternative and complementary reproductive tactics. *American Naturalist* 115:223–246.
Conry, P. J. 1988. High nest predation by Brown Tree Snakes on Guam. *Condor* 90:478–482.
Coon, R. A., J. D. Nichols, and H. F. Percival. 1981. Importance of structural stability to success of Mourning Dove nests. *Auk* 98:389–391.
Cowan, J. B. 1952. (Full reference in list for chapter 3.)
Duff, C. V. 1951. Mourning Doves raise triplets. *Condor* 53:258–259.
ffrench, R. 1973. *A Guide to the Birds of Trinidad and Tobago.* Wynnewood, Pa.: Livingston Publishing Co.
Friedmann, H., and L. F. Kiff. 1985. The parasitic cowbirds and their hosts. *Proceedings of the Western Foundation of Vertebrate Zoology* 2:225–302.
Hanson, H. C., and C. W. Kossack. 1963. (Full reference in list for chapters 5, 6, and 7.)
McClure, H. E. 1942. Mourning Dove production in southwestern Iowa. *Auk* 59:64–75.
Marchant, S. 1960. (Full reference in list for chapters 5, 6, and 7.)
Murton, R. K. 1965. (Full reference in list for chapter 1.)
Murton, R. K., R. J. P. Thearle, and J. Thompson. 1972. Ecological studies of the Feral Pigeon *Columba livia* var. I: Population, breeding biology, and methods of control. *Journal of Applied Ecology* 9:835–874.
Murton, R. K., N. J. Westwood, and A. J. Isaacson. 1974. Factors affecting egg-weight, body-weight and moult of the Wood Pigeon *Columba palumbus. Ibis* 116:52–73.
Nichols, J. D., H. F. Percival, R. A. Coon, M. J. Conroy, G. L. Hensler, and J. E. Hines. 1984. Observer visitation frequency and success of Mourning Dove nests: A field experiment. *Auk* 101:398–402.
Passmore, M. F. 1984. Reproduction by juvenile Common Ground-Doves in South Texas. *Wilson Bulletin* 96:241–248.
Skutch, A. F. 1981. (Full reference in list for chapters 5, 6, and 7.)
Van Someren, V. G. L. 1956. (Full reference in list for chapters 5, 6, and 7.)
Westmoreland, D., and L. B. Best. 1985. Effects of researcher disturbance on Mourning Dove nesting success. *Auk* 102:774–780.
Westmoreland, D., and L. B. Best. 1986. Incubation continuity and the advantage of cryptic egg coloration to Mourning Doves. *Wilson Bulletin* 98:297–300.
Westmoreland, D., and L. B. Best. 1987. What limits Mourning Doves to a clutch of two eggs? *Condor* 89:486–493.

Westmoreland, D., L. B. Best, and D. E. Blockstein. 1986. Multiple brooding as a reproductive strategy: Time conserving traits in Mourning Doves. *Auk* 103:196–203.
Wiley, J. W., and B. N. Wiley. 1979. (Full reference in list for chapter 2.)

Chapter 9 Pigeons and Man

Bucher, E. H. 1970. (Full reference in list for chapter 3.)
Bucher, E. H. 1974. *Bases ecológicas para el control de la Paloma Torcaza.* Publicación No. 4. Córdoba, Argentina: Centro de Zoología Aplicada, Universidad Nacional de Córdoba.
Cottam, C., and J. B. Trefethen, eds. 1968. (Full reference in list for chapter 1.)
Goodwin, D. 1978. *Birds of Man's World.* Ithaca and London: British Museum (Natural History) and Cornell University Press.
Leahy, C. 1982. *The Birdwatcher's Companion: An Encyclopedic Handbook of North American Birdlife.* New York: Hill and Wang.
Levi, W. M. 1951. *The Pigeon.* Columbia, S.C.: R. L. Bryan Co.
Murton, R. K. 1965. (Full reference in list for chapter 1.)
Murton, R. K., C. F. B. Coombs, and R. J. P. Thearle. 1972. Ecological studies of the Feral Pigeon *Columba livia* var. II: Flock behaviour and social organization. *Journal of Applied Ecology* 9:875–889.
Murton, R. K., R. J. P. Thearle, and J. Thompson. 1972. (Full reference in list for chapter 8.)
Williams, T. 1985. The quick metamorphosis of Indiana's doves. *Audubon Magazine* 87:38–45 (and correspondence in subsequent issues).

Chapters 10 and 11 Homing Pigeons and How Pigeons Find Their Way

Able, K. P. 1980. Mechanisms of orientation, navigation, and homing. In *Animal Migration, Orientation, and Navigation,* edited by S. Gauthreaux. New York: Academic Press.
Alexander, J. R., and W. T. Keeton. 1972. The effect of directional training on initial orientation in pigeons. *Auk* 89:280–298.
Alexander, J. R., and W. T. Keeton. 1974. Clock-shifting effect on initial orientation of pigeons. *Auk* 91:370–374.
Bingman, V. P. 1987. Earth's magnetism and the nocturnal orientation of migratory European Robins. *Auk* 104:523–525.
Emlen, S. T. 1967. Migratory orientation in the Indigo Bunting, *Passerina cyanea.* Part 1: Evidence for the use of celestial clues. Part 2: Mechanism of celestial orientation. *Auk* 84:309–342, 463–489.
Emlen, S. T. 1970. Celestial rotation: Its importance in the development of migratory orientation. *Science* 170:1198–1201.
Herbert, R. A., and K. G. S. Herbert. 1965. Behavior of Peregrine Falcons in the New York City region. *Auk* 82:62–94.
Hitchcock, H. B. 1955. Homing flights and orientation of pigeons. *Auk* 72:355–373.
Keeton, W. T. 1970. Comparative orientational and homing performances of single pigeons and small flocks. *Auk* 87:797–799.

Levi, W. M. 1951. (Full reference in list for chapter 9.)
Lipp, H. P. 1983. Nocturnal homing in pigeons. *Comparative Biochemistry and Physiology A* 76:743–749.
Matthews, G. V. T. 1968. *Bird Navigation.* 2d ed. Cambridge, England: Cambridge University Press.
Matthews, G. V. T. 1985. "Homing Pigeon" and "Navigation." In *A Dictionary of Birds,* edited by B. Campbell and E. Lack. Calton, England: T. and A. D. Poyser.
Maubouche, S. 1986. Frontiers of medicine: On a wing and a test tube. *Washington Post,* February 2.
Merkel, F. W., and W. Wiltschko. 1965. Magnetismus und Richtungsfinden zugunruhiger Rotkehlchen (*Erithacus rubecula*). *Vogelwarte* 23:71–77.
Nicol, J. A. C. 1945. The homing ability of the carrier pigeon: Its value in warfare. *Auk* 62:286–298.
Papi, F. 1986. Pigeon navigation: Solved problems and open questions. *Monitore Zoologico Italiano* (N.S.) 20:471–517.
Schmidt-Koenig, K. 1960. Internal clocks and homing. *Cold Spring Harbor Symposiums on Quantitative Biology* 25:389–393.
Schmidt-Koenig, K. 1987. Bird navigation: Has olfactory orientation solved the problem? *Quarterly Review of Biology* 62:31–47.
Walcott, C., and K. Schmidt-Koenig. 1973. The effect on pigeon homing of anesthesia during displacement. *Auk* 90:281–286.
Waldvogel, J. A. 1987. Olfactory navigation in homing pigeons: Are the current models atmospherically realistic? *Auk* 104:369–379.

Chapter 12 Darwin's Pigeons

Darwin, C. 1900. *The Variation of Animals and Plants under Domestication.* New York: Appleton.

Index

The locations of illustrations are shown in **boldface** type.